*Routledge Revivals*

---

# Modelling Housing Market Search

Originally published in 1982, this book contains research in the area of econometric modelling in the housing market, including that which has extended to the use of search models. The subjects covered include the importance of racial differences, spatial aspects of residential search and information provision and its effect on the behaviour of the buyers. The combination of careful analytic modelling, empirical testing and speculative discussions of the role of agents in the search process provides an innovative and imaginative approach to the interesting problems of understanding the individual behaviour in complex contexts such as the urban housing market.

# Modelling Housing Market Search

Edited by W. A. V. Clark

Routledge
Taylor & Francis Group

First published in 1982
by Croom Helm Ltd

This edition first published in 2021 by Routledge
2 Park Square, Milton Park, Abingdon, Oxon, OX14 4RN
and by Routledge
605 Third Avenue, New York, NY 10158

*Routledge is an imprint of the Taylor & Francis Group, an informa business*



**Publisher's Note**
The publisher has gone to great lengths to ensure the quality of this reprint but points out that some imperfections in the original copies may be apparent.

**Disclaimer**
The publisher has made every effort to trace copyright holders and welcomes correspondence from those they have been unable to contact.
A Library of Congress record exists at LCCN:82000872

ISBN 13: 978-1-032-02143-0 (hbk)
ISBN 13: 978-1-003-18208-5 (ebk)

DOI: 10.4324/9781003182085

# Modelling Housing Market Search

Edited by W.A.V. Clark

CROOM HELM
London & Canberra

Croom Helm Ltd, 2-10 St John's Road, London SW11

British Library Cataloguing in Publication Data

Clark, William A.V.
Modelling housing market search.
1. Housing–Mathematical models
I. Title
339.4'83635'0724 HD7287.5
ISBN 0-7099-0726-5

Printed and bound in Great Britain by
Biddles Ltd. Guildford and King's Lynn

# CONTENTS

# PREFACE

The papers in this volume were originally prepared for a seminar on housing market search held at the University of California, Los Angeles, during May of 1980. The papers presented at that seminar have since been substantially revised and extended and additional papers on housing search have been added to the collection. The seminar was sponsored by the Institute for Social Science Research at U.C.L.A. and held in association with a graduate course on migration and mobility. It was designed to bring together geographers, economists, sociologists, and planners to discuss the fast-developing field of residential search. The emphases included both theoretical and empirical aspects of search in the housing market as well as the policy implications of studies of search.

Several geographers and demographers in Southern California and the sabbatical visit of Eric Moore from Queens University provided a nucleus of individuals interested in residential mobility and search. The group was augmented with invitations to John Goodman and Francis Cronin of the Urban Institute, Duncan Maclennan of the Center for Urban and Regional Research at the University of Glasgow, James Huff from the Department of Geography at the University of Illinois, and Risa Palm from the University of Colorado. This combination of disciplinary approaches allowed a rich interchange of both theoretical and empirical perspectives.

The papers include an extensive review of search models and their applications to search in the housing market and several conceptual and theoretical models of search, including the formulation of one of the first explicitly spatial models of search. Other papers focused on the role of information in search and possible policy responses to housing market search. One of the major purposes of this volume is to provide a readily accessible collection of major works which are representative of the ongoing research in this rapidly developing area.

The conference would not have been possible without the support of the Institute for Social Science Research and I would like to thank the director, Howard Freeman, for making available funds for research on residential mobility and migration. I would also like to thank the Department of Geography at U.C.L.A., especially Noel Diaz for cartographic services in preparing the manuscript.

Los Angeles, California — William Clark

# Part I

# MODELS OF SEARCH BEHAVIOR

The chapters in this book have been divided into two broad sections. The first of these sections focuses on models of search behavior. The second is concerned with the nature of information and its influence on search behavior. The chapters in the first part of the book are organized to review the general nature of search models, and particularly, their application to search in the housing market, to present a conceptual approach to housing search, and specific analytic models of the search for housing. The papers represent general conceptual and specific analytic models and empirical tests of the models. They are representative of the latest thinking on applied residential search models in geography and economics.

While the work by economists on job search stimulated the development of a variety of econometric models of search behavior, most of this work focused on stopping rule models and has had only limited application to search in the housing market. The economic research emphasized equilibrium analysis and the clearing conditions in the market. Recently, specific studies of housing market search have attempted to develop more realistic models of search behavior. The first chapter organizes previous analyses of housing market search into six categories, two of which are concerned with disequilibrium models of search, another focuses on stopping rule models, and the other three focus on the analysis of institutional constraints on search, the role of information, and specific studies of spatial search in the housing market. The chapter argues that much of the research has been cross-sectional and emphasized either utility maximizing approaches or institutional intervention. However, the most recent research is clearly focused on the central role of information in search behavior. As an overview the review sets up a basis for the more specific chapters which follow. Kevin McCarthy outlines a conceptual model of housing search and mobility, Gavin Wood and Duncan Maclennan discuss search adjustment in the local housing market of Glasgow, Frank Cronin analyses racial differences in the search for housing and Jim Huff models the spatial aspects of residential search.

The second chapter outlines a three-stage model of the search process in which there is a distinction between the decision to move, the decision to search, and the housing choice itself. McCarthy specifies an analytic formulation for each of these stages which emphasizes local mobility as the vehicle for housing consumption adjustment. One of the major elements of this conceptual model is the role of transaction costs in search and the discounts that people get from various intensities of search. Although a complete test of the model has not yet been concluded, there is preliminary evidence to indicate that there is considerable variation amongst households (in their search behavior), that supply conditions are likely to affect search processes, and finally, that search procedures appear to affect outcomes, although not necessarily in the same way for owners and renters.

Wood and Maclennan contend that the static neo-classical models of consumer demand do not represent a plausible framework in which to analyze housing choice situations. They outline the relevant features of the housing market and then establish that search activities are an important component of the process of purchase. Given this premise, they develop a model of "the search adjustment process" in which search costs and information are central components of search. They investigate both search by students in the rental housing market and entry into the privately owned housing market. One of the most interesting parts of the model and its empirical test in the Glasgow housing market is the realization that the investigation of search in the housing market may be as relevant in a controlled housing market as it is in a "free" market.

Chapter 4 is an illustration of the classic disequilibrium approach to search. Cronin uses this approach to analyze the effectiveness of search by minority households, including the differences in information services, modes of transportation used in search, the length of search, and the number of neighborhoods searched. The chapter is an excellent example of the disequilibrium approaches that have been the mainstay of models of search behavior in the housing market. The results confirm differences (for minority and nonminority households) that are not so much in the ends obtained from the search process, but rather, from quite different strategies involved in search behavior.

The final chapter in the first part of the book is a specific investigation of spatial search. As Huff argues, one of the basic stimuli for investigating spatial search relates to the integral link between where people are looking and where people ultimately relocate. And, if there is some stability in the patterns of household search then these may provide a clue to ultimate population relocation patterns. Huff develops what he calls an area-based search model in which an area is first selected for search and second vacancies within the area are targeted. The model assumes that the household uses one or more sources to acquire indirect information on selected attributes of vacancies -- certain vacancies are rejected solely on the basis of the indirect information, while others are considered to be viable possibilities. The household proceeds to visit vacancies in the possibility set until a new residence is found or the household stops searching. Members of the possibility set which are located in areas of the city with relatively low expected search costs are assumed to have a higher probability of being identified for visits than members located in higher search cost areas. The formal model of the search process is shown to be a first order Markov chain, where the probability of searching in a given area is a function of the location of the last vacancy seen by the household and the relative concentration of possible acceptable vacancies.

These chapters illustrate two important points. First, they represent attempts to take what we know or suspect about the spatial aspects of residential search, to express these ideas in formal terms and to test the models against hypothetical or real world data. It is this attempt to both formalize and test models of search which distinguishes these contributions to residential search. The papers are set in a real world framework and represent attempts to bring mobility and the housing market into a closer association. The second point that is established by the international comparative framework of these papers is that the constraints versus choice arguments which have been so common in the past half dozen years may be overstated. The Wood and Maclennan paper suggests that search is an important element even in a constrained housing market. In their study of the Glasgow market (in which rent controls were present throughout the study period) they provide convincing evidence that the allocation mechanism of apartments or flats was based on the search procedures of individuals, rather than any spatial rationing by a rent bidding process. This emphasizes the role of choice even in situations in which there is a controlled market. It is quite clear that in European housing markets, the patterns of residential location and the decisions of individual households as they move in the city are constrained by a variety of individual and institutional forces. It is this difference in context between North American and European markets which has engendered a continuing debate between those individuals who argue that the research focus should be on the constraints alone, as against those individuals who consider the choice dimension to be more important and who view constraints as simply limits on choices.

# 1

# A REVIEW OF SEARCH MODELS AND THEIR APPLICATION TO SEARCH IN THE HOUSING MARKET

**W. A. V. Clark and Robin Flowerdew**

Although the interest by social scientists in search behavior is little more than twenty years old, there is now a substantial literature in several disciplines, including psychology, marketing, economics, and geography. This literature includes both theoretical models of search under differing conditions and empirical work on search processes based on experimental and survey techniques. The literature is sufficiently diffuse and has such distinct disciplinary foci to make it worthwhile surveying both the empirical and theoretical literature and to draw out of that literature the elements which are most important for the adaptation of the theoretical work for further developments of models of search behavior in the housing market.

Studies of search behavior have been a natural outgrowth of the study of how individuals make decisions. Edwards (1954) initially outlined a basic structure for the investigation of decision-making in which he was concerned to find out how people evaluate the utilities of various objects, how they judge the probabilities of events, how these probabilities change with information, and how the probabilities and utilities are combined to control decisions. Most studies of search in the social sciences can still be set within this framework. The framework will not be reviewed here, but it is important to emphasize that studies of search are part of the wider context of research on individual decision-making. Indeed, one of the main reasons for the study of search is its effect on the decision-making process.

In a complex society, individual choice and decision-making processes (including the process of search for alternatives) are increasingly important and have implications in a wide range of situations. In an earlier time, job choice was largely determined by parental occupations or by opportunities available locally. Now the increasing fluidity of modern society has created a situation in which job search and job choice are critical to the functioning of the system. Similarly, an increasingly complex consumer society presents people with a multitude of choices whose outcomes are of considerable importance to the economy. Consumer decisions with respect to housing are critical both for the individual and the housing market.


DOI: 10.4324/9781003182085-2

This chapter begins with an analysis of the contributions to the theory of search from a disciplinary perspective, reviewing contributions made in psychology, marketing, economics, and geography. Second, some important concepts in theoretical and empirical discussions of search are identified. Third, these basic notions are applied to the housing market context, and brief reviews are given of papers which have developed models of residential search and choice. Finally, we present some speculations on the implications of search models for the development of policy on housing and residential mobility.

## DISCIPLINARY PERSPECTIVES

Initially, approaches to the study of search were defined by disciplinary perspectives, but now the particular contributions of each discipline have been at least loosely integrated. From psychology has come an emphasis on information and decision-making; from economics, an emphasis on optimality; from marketing, an emphasis on personal characteristics and individual differences; and from geography, an emphasis on the spatial context. Concern with search in a particular empirical context has stimulated the integration of these disciplinary approaches. Search in the housing market is a complicated search process, and an understanding of this process necessitates the use of ideas from psychology, marketing, economics, and geography. It is useful to understand the literature in each of these disciplinary traditions and their contributions to the way in which models of search in the housing market have developed.

### Psychology

In psychological research, questions of search are usually considered in the context of questions of choice. Choice among a set of alternatives implies the existence of a search process to locate and investigate the alternatives, but the psychology literature focuses on how the alternatives are evaluated rather than on how they are identified. Psychologists have especially stressed the role of information in the decision-making process, and have discussed the way in which information is incorporated into the evaluation of alternatives.

There are several reviews of work in psychology on information processing in decision-making, including papers by Slovic and Lichtenstein (1971) and Slovic, Fischhoff and Lichtenstein (1977). These papers distinguish two approaches to the analysis of information use. The first, integration theory uses multiple regression and analysis of variance to analyze the way in which information is used in judgments. The second, Bayesian analysis, emphasizes the role of subjective probabilities and the revision of opinion in the light of new information. These studies emphasize the amount of information collected, the form in which it is collected, and the way in which it is processed. Dynamic decision-making (Rapoport, 1975; Kleiter, 1975) emphasizes studies of how information modifies the beliefs of individuals or households and, as a result, causes changes in their actions. The way in which information is acquired and processed may affect the final decisions that are made. Rapoport and his co-workers (Kahan, Rapoport and Jones, 1967; Rapoport, Lissitz and McAllister, 1972) have conducted experiments on the strategy adopted by subjects who are faced with a decision as to when to stop collecting information and make a choice. Brickman's work (1972) makes it clear that a subject's behavior tends to be suboptimal when the alternatives increase or decrease systematically in value. Other studies have examined the way in which resource limitations or other constraints affect the processing and use of information (Levine, Samet and Brahlek, 1975).

Most of the psychology literature on choice is primarily concerned with individual judgments. For example, the information integration approach is specifically concerned with how individuals put information together and the relative weighting that they give to the stimulus parameters, that is, to the variables that go into information development and thus influence the subjective values of information arguments that are used by individuals (Anderson, 1968; 1974). Studies in psychology have emphasized the role of information and the way in which information is used in decision-making rather than providing approaches to search and information use in sequential decision-making. The recent interest in information collection in sequential decision processes, however, places more emphasis on the role of search in the decision process.

## Marketing

Research in marketing has paralleled investigations in psychology. Many studies in the marketing field have focused on tests of psychological models as applied to consumer choice behavior. These studies have employed a variety of experimental structures and games which can be linked to particular aspects of information use in decision-making. There have been studies of information acquisition, particularly with respect to the amount and source of information used, and to the impact of time pressure and distraction on the process of information acquisition. Survey methods and experimental games have dominated the marketing literature. Bettman (1976; 1979) analyzes the major problems of each of the approaches, and includes a thorough analysis of the marketing research literature, particularly those experimental approaches concerned with the nature and amount of information used by consumers in decision-making.

## Economics

In the past twenty years, the most important theoretical advances in the modelling of search have taken place in economics, mostly in the context of job search. Stigler (1961) pointed out that the limited amount of information available to job seekers would affect their optimal behavior. McCall (1970) expanded Stigler's approach, and several economists have extended these job search models and explored their implications for the economy as a whole (Rothschild, 1973; Pissarides, 1976; Lippman and McCall, 1979). A detailed review of job search models is provided by Lippman and McCall (1976).

Most of these studies are based on stopping rule models. Such models concern the decision which must be made between stopping search to accept a job offer already located and incurring the costs of further search. Some models allow the decision-maker to revise his expectations about wage levels he is likely to be offered on the basis of the wages of job offers already located. In other cases (e.g. Salop, 1973), hierarchical models are developed in which the searcher can apply for jobs with any of several firms, for each of which he can estimate the expected wage level.

An important aspect of these studies is their focus on the influence of information and uncertainty on optimal search procedures. Articles by Ioannides (1975), Kohn and Shavell (1974), Rothschild (1973) and Telser (1973) extend the theory of search and emphasize the description of search behavior as sequential decision-making under uncertainty. Rothschild (1973) utilizes the notion that consumers learn about the probability distribution of prices while they search from it. Thus, he emphasizes the role of search for learning as well as in locating alternatives. And, in particular, he is able to develop optimal search rules from unknown distributions and to show that these optimal search rules have the same qualitative properties as optimal rules from known distributions (under fairly strict assumptions). Rothschild's approach emphasizes the learning and updating of prior beliefs in a Bayesian manner as an individual searches. Telser (1973) also stresses the role of prior information. If the searcher knows the distribution of the alternatives of interest (say, jobs), then he can search optimally. Otherwise he must postpone his decision and gather information. In general, the economists have developed search models within the utility maximization framework and have attempted to link these models to aggregate characteristics such as unemployment and the Phillips curve.

Only in the recent past have economists addressed the components of search in the housing market. Ioannides (1975) has come closest to a discussion of the behavioral components of search. He attempts to integrate the search process into the market structure and focuses specifically on some behavioral rules which market participants are presumed to follow.

> The state of the market is described by probability distribution functions of sellers' asking prices and buyers' reservation prices and the number of prospective buyers and sellers in the market. Buyers' (sellers') search is modelled as sampling at random times from the distribution of asking (reservation) prices. The process of contacts is described by a generalized Poisson whose intensity is a function of market conditions. Market participants search expecting stationary market conditions. Equilibrium adjustment of the distributions of asking and reservation prices is described by modelling the market participants' "flowing through" the price range characterizing the market. Some make contacts and others do not, and some of the contacts made do not lead to transactions. Buyers and sellers set finite horizons for their search. If, at the end of the time-horizon, a seller has not sold or a buyer has not bought, they leave the market (Ioannides, 1975:260-261).

Although the emphasis in such work has been theoretical, some empirical work on job search has been conducted (Granovetter, 1974; Kahn, 1978; Melnik and Saks, 1977; Schiller, 1975). Empirical work has also been carried out on search in the rental housing market (Maclennan, 1979), and Chinloy (1980) has developed and tested a search model of the market for single-family homes.

## Geography

The main emphasis in geographic studies of search has been on spatial factors, although several studies have used non-spatial search models to illuminate decision-making processes of interest in geography. Spatial search was discussed by Gould (1966), but it is only recently that geographers have given systematic attention to the topic. Silk (1971) reviewed a series of studies on search behavior from the behavioral sciences, and Schneider (1975) developed some models for spatial search in urban areas, mainly concerned with search as a means of locating a target, rather than search as the identification of alternatives amongst which to choose.

The development of interest in the late 1960s in individual spatial behavior was concentrated around a few decision-making problems, including residential mobility and shopping behavior. The concept of awareness space was developed to describe that area of a city in which an individual would be aware of possible alternatives (Wolpert, 1965). Work on urban contact fields emphasized the spatial variation in the probability that particular places will be within someone's awareness space (Moore and Brown, 1970; Moore, 1970). Other work suggested that the shape of most people's awareness space would result in a tendency for search for new housing to be concentrated in a sector extending outwards from the present home in the same direction from the central business district (Adams, 1969).

Several studies have investigated the behavior of households searching for new housing. Barrett (1973), for example, collected information in Toronto about the temporal duration of the search process, the number of houses examined, and their locations. He also studied spatial variation in the use of information channels, finding differences between the behavior of people living in high-income and low-income residential areas, findings in accord with those of Herbert (1973) in South Wales. Palm (1976a; 1976b) has investigated the role of real estate agents in influencing the search processes of home-seekers. A real estate agent may encourage clients to search in areas which he or she thinks are most likely to have suitable houses. Palm shows that realtors' advice may be guided by their own perception and knowledge of a city, and hence biased towards those areas they know best themselves. Hartshorn (1980) illustrates the way newspaper advertisers attempt to influence housing search by providing social cues to the properties on offer as well as architectural and economic data.

Stopping rule models were reviewed by Flowerdew (1976), and have been used in studies of shopping behavior (Hay and Johnston, 1979), as well as residential mobility (Smith, Clark, Huff, and Shapiro, 1979; Meyer, 1980) and inter-urban migration (Miron, 1978). Attempts have been made to test these models, using both experimental methods (Meyer, 1980; Phipps, 1978) and survey work (Smith and Clark, 1982; Clark and Smith, 1982). The application of these models has often not been explicitly spatial. There has been a tendency to treat location as merely one attribute of the alternatives to be considered, or as a way in which a household's search strategy may be organized (Smith et al., 1979), although Huff (1981) has directed attention back to the spatial distribution of vacancies inspected.

At least in comparison with psychology and economics, work on search in geography has been fairly eclectic, drawing concepts and ideas from several other disciplines. It has also been characterized by continuing attempts to link the theoretical and the empirical, and to apply abstract and simplified models to the real world complications and constraints of actual search experience.

## CONCEPTS OF SEARCH

There are a number of concepts discussed in the literature on search behavior which are particularly relevant to the study of search in the housing market. A proper understanding of housing search must be based on these concepts. First, search is a goal-directed activity. It is undertaken by people who have some idea of what they want, and who can organize their search activities in a way which they think will be effective. Their decision should not therefore be regarded as a choice between alternatives randomly selected from the total set of dwelling units. Instead, the search procedure will concentrate on the methods of acquiring information that are thought most likely to generate the right kind of alternatives.

Second, search involves a complex process of information gathering. It is not just a matter of locating alternatives; it also involves collecting substantial amounts of information about them to be used in the process of decision-making. A related point is that search may be regarded as hierarchical; before specific dwelling units are located, it may be necessary to search for, and choose between, alternative information channels, such as alternative real estate companies, local newspapers, or display boards. Often, decisions must be made about which neighborhoods should be selected for search before individual dwelling units are located and assessed.

Third, the search process takes place under uncertainty. The searcher is never aware of the complete set of opportunities, and does not know if the use of a different information channel might bring to light better alternatives than he or she has already located. In addition, there may be uncertainty about some attributes of a particular alternative. This uncertainty may be resolved at the cost of further investigation, or may be impossible to allay until the decision has been made. The contribution of economists to the theory of decision-making under risk and uncertainty is fundamental to an understanding of the impact of these factors on the search process. Other important ideas in this area are the concepts of the expected utility of search and of the existence of a probability distribution on the utility of dwelling units in a neighborhood.

Fourth, there must be some way to determine when the search process can be ended and a choice made. This decision is dependent on the adequacy of the best alternative located, on the searcher's assessment of whether a better alternative can easily be found, and on the costs of continuing search. These costs can include any financial outlay that may be involved, the time, effort, or frustration of further search, opportunity costs that may be incurred, or the possibility of losing a desirable alternative to somebody else by failing to accept it immediately. Many stopping rule models have been developed, principally in economics and operations research, to apply to different sets of search objectives, to different amounts of information about unexamined alternatives, and to different types of search costs.

Finally, the search process takes place within a set of constraints. These may affect the money or time available to searchers, or the access they have to information sources. Flowerdew (1978) has examined the relevance of different types of time constraints to residential search. The operation of information channels may constrain search substantially: real estate offices may supply information only for dwelling units of a particular type or in a particular area; they may also steer searchers of a particular race, class, or family type to one type of accommodation or area, making it hard for them to find information about other types. In addition to constraints on search, there may be important constraints on the availability of some opportunities to some searchers; in addition to racial, cultural, and life-style factors, the availability of financing may be vital in the process of buying a house. Some would-be purchasers may have to coordinate a search for financing with their search for a home.

## SEARCH IN THE HOUSING MARKET

Much work on residential mobility, especially in geography and sociology, has been indebted to Rossi's landmark study, Why Families Move (1955). Rossi presented some empirical findings on information channels used in housing search, and his lead has been followed by many others (see Clark and Smith (1979) for a review). Only in the last few years, however, has there been any development of theoretical models of housing search, or much attempt to link the empirical material to these theoretical ideas. This recent literature encompasses some quite distinct approaches and themes. It can be classified into six broad categories. The first group of studies which utilize a disequilibrium model of housing adjustment are only in the broadest sense concerned with search. The initial disequilibrium model (Goodman, 1976) has been extended by Hanushek and Quigley (1978) and tested by Weinberg et al. (1979) and Cronin (1979a). The main focus of these models is on the benefits of moving. A second group of studies have attempted to extend the disequilibrium approach with a specific treatment of search (Weinberg et al., 1977; Cronin, 1979b). A third approach emphasizes the role of institutions (such as the real estate industry) in housing search and the effect of discrimination (Palm, 1976a; Smith and Mertz, 1980; Courant, 1978). A fourth approach to search in the housing market has concentrated on the role of information in search behavior (Clark and Smith, 1979). Fifth, there have been several studies attempting to adapt stopping rule models to the housing market (Friesz and Hall, 1978; Smith et al., 1979). Lastly, a few studies have concentrated on specifically spatial aspects of search (Huff, 1981).

### Disequilibrium Models

The disequilibrium model and its variations were developed as alternatives to the traditional sociological and geographical models of residential mobility. The proponents of disequilibrium models emphasize the link between mobility and economic decision-making based on the benefits and costs of moving. Empirical analysis, however, indicates that economic models generally do not have good predictive power. Even where they are significant, variations in the economic factors do not lead to large changes in the rates of search or mobility. The disequilibrium models share a similar structure -- they apply to individual households, the household is assumed to make a decision on mobility based on the economic benefits and costs of moving, and a logit or probit model is used to link the measures of benefits and costs to the rate of search or moving.

In an initial specification of the disequilibrium model, Goodman (1976) expressed the probability that a household will move at least once during a given time period as a linear function of benefits and costs of moving. Housing consumption is expressed as a vector of four factors: expenditure, size, quality, and workplace accessibility. The deviations of the factors of the current consumption from those of the optimal consumption are assumed to be the potential benefits of moving. Following traditional economic theory, the optimal consumption of housing services, hence also expenditure, is derived from maximizing household utility subject to a budget constraint and is computed from an estimated demand function for housing. The optimal size, quality, and accessibility are arbitrarily defined. The costs of moving include the transaction cost and the psychological cost of leaving the present unit and neighborhood. The out-of-pocket moving cost is assumed constant for all households, hence ignored in the analysis. The transaction and psychological costs are proxied by a vector $\underline{Z}$ of tenure, year of last move, age of household head, and household life cycle stage.

In the Hanushek and Quigley formulation, household mobility (probability of search, probability of move) is assumed to be a probit function of the difference between actual housing consumption at time t and equilibrium consumption demanded at time t+1, divided by equilibrium consumption demanded at time t:

$$M_t = f\left(\frac{H^d_{t+1} - H_t}{H^d_t}\right) \quad (1)$$

where

$M_t$ = probability of move (or of search) between t and t+1.
$H_t$ = actual consumption of housing services, measured by monthly rent, adjusted for contract terms and length of tenure.
$H^d_t$ = equilibrium housing consumption demanded at time t.

All transaction costs (search and moving costs) are assumed to be randomly distributed, independent of the demand disequilibrium, and are therefore ignored in the model. Housing demand is assumed to be a linear function of a household's socio-economic characteristics, including household income, household size, and race and age of the household head. The model is also modified in order to separate the effects of a change in equilibrium housing demand between t and t+1, and current disequilibrium in consumption.

Cronin (1979a) explores two alternative measures of the benefits of moving. The first is disequilibrium in housing consumption, as defined by Hanushek and Quigley (1978), but with a different approach to estimating the equilibrium demand function. The second measure is income equivalent variation (IEV), which is the amount of additional income necessary to make the household as well off with its current consumption of housing as it would be with its equilibrium consumption. "As well off" is defined by the same value of cardinal utility. Both types of benefit measures are estimated using two different forms of utility functions: Cobb-Douglas and Stone-Geary.

Weinberg, Friedman, and Mayo (1979) analyze the effect of rent rebates on households' residential mobility. In the empirical analysis (based on data from the Housing Allowance Demand Experiment) the authors use a logit model of housing search and mobility, and include in the set of independent variables an estimated change in consumer surplus induced by the rent rebate. Household mobility is assumed to be a logit function of benefits and costs of moving. Estimated costs include: expected out-of-pocket moving costs, expected length of search, and psychological cost, proxied by current tenure discount. The potential benefit of changing from the actual (non-optimal) housing to the equilibrium (optimal) housing is measured by the increase in consumer surplus due to the change. Consumer surplus is calculated from an assumed double logarithmic demand function for housing services and is expressed as a function of current expenditure on housing, $p_H H_o$, equilibrium expenditure on housing, $p_H H^*$, and price elasticity of demand for housing, b.

The principal difference among the models lies in the method used to calculate the benefits of moving. Hanushek and Quigley (1978) use the difference between the actual and equilibrium levels of housing consumption, expressed as a fraction of current equilibrium demand. Weinberg, Friedman, and Mayo (1979) estimate the potential change in consumer surplus due to moving, calculated from an assumed demand function for housing services. Cronin (1979a) computes the income equivalent variation of a move from actual to equilibrium consumption of housing, making specific assumptions regarding the form of the household utility function. Estimated change in consumer surplus is in fact an approximation of income equivalent variation, hence, the models of Weinberg et al. and Cronin are somewhat alike. They differ, however, from Hanushek and Quigley in recognizing the possibility that a household can substitute other consumption goods for housing.

## Disequilibrium Models of Search

Both Cronin (1979b) and Weinberg et al. (1977) examine search within the disequilibrium framework. Cronin uses an economic measure of benefits of moving to explain the rate and characteristics of search, and the effect of race on search. The model used is a modified version of the Stone-Geary, IEV model developed in the earlier paper (Cronin, 1979a). The probability that a household undertakes residential search within a twelve-month period is assumed to be a logit function of benefits, costs, psychological factors, past mobility, and, in addition, race. Cronin examines a number of indicators of search effort, such as number of days searched, number of neighborhoods searched, and number of dwelling units searched. He estimates separate multiple regressions using the indicators of search effort as dependent variables and economic variables and household characteristics as independent variables. The results indicate that IEV affects search behavior in an expected way: the greater the IEV, the shorter the search period and the more intense the search.

The mean values of some of the variables related to search differ significantly between minority and non-minority households. Average IEV for minority households is $524/yr., but $156/yr. for non-minorities. The average number of search days is over twice as many for minority as non-minority households. On the average, minority households appear to have greater potential benefits from moving, but also to face greater costs of search than the non-minority households. The overall probability of a move does not differ very much between the races, but minority and non-minority households follow different strategies during search. Significantly more non-minority households use newspapers than minority households and minority households tend to use vacancy signs more often than the non-minorities. Fewer minority households use their own automobile for search, relying more on taxis and public transportation than the non-minority households. There are also significant differences between the types of neighborhoods searched by minority and non-minority households, indicating that households may anticipate discrimination and avoid search in some neighborhoods. (It seems difficult to say, however, whether the households are anticipating discrimination or simply availability of appropriate types of housing units.)

Weinberg et al. (1977) use the disequilibrium framework to examine a household's rate of search and moving. The logit model estimates the probability of search primarily on the basis of socio-economic characteristics of the household together with previous mobility, race, reported satisfaction with the housing unit and the neighborhood, and a dummy variable separating experimental and control households. The most significant coefficients are related to previous mobility, age of household head, dissatisfaction with housing unit, and dissatisfaction with neighborhood.

A logit model with the same independent variables as those applied to the probability of search is estimated for the household's decision to move, given search, but the fit of the model is quite poor. They conclude that the effort expended by searchers who move is not significantly different from that of searchers who do not move. Hence, the failure of some searchers to move is apparently due not to lack of effort but to problems encountered while searching. They suggest that the main barriers encountered during search include a lack of knowledge about available housing, problems with transportation, problems related to children, and expected financial difficulty. Although financial difficulty is the only barrier that could be alleviated by the housing allowance, financial assistance did not appear to make moving more likely.

## Institutional Factors and Discrimination

Some of the same issues have been discussed outside the disequilibrium framework, in terms of the discriminatory effects of real estate agents and other information channels. Courant (1978) and Yinger (1978) have developed theoretical models of the real estate impact on search and Palm (1976a) has studied the impact of real estate agents on the "spatial bias" of house choices. (See Clark and Smith, 1979, for a review.) Smith and Mertz (1980) also studied the potential impact of realty agents on the ordering of vacancies seen by a client.

Another important analysis of the effects of discrimination on search behavior was carried out by McCarthy (1979) as part of the Housing Allowance Supply Experiment in Brown County, Wisconsin, and St. Joseph County, Indiana. The two major issues addressed by McCarthy are: (1) how and why search strategies vary with household characteristics, and (2) the effect of the choice of strategy on the household's ability to locate a bargain (households could either minimize search cost or maximize the likelihood of obtaining the best alternative). The choice of strategy also involves the choice of information source and the amount of effort to be expended. The data examined are limited to renter households who had moved into their current dwellings within the preceding five years from elsewhere in the same county. The sources of information used in search include newspapers, friends and relatives, walking and driving, and real estate agents. The data appear to confirm a minimum cost strategy among renters.

The effect of search strategy on locating "bargains" is analyzed by a two-step process. A hedonic price equation, estimated by previous work, is used to determine the expected price of a rental unit. The residual between the actual and the expected prices indicates the amount of discount or premium associated with the unit. The residuals are then regressed against a vector of search and household variables. The results of this analysis indicate that the renter encountering problems during search is likely to pay a premium for the unit he ultimately selects. At the same time, a low intensity search strategy, relying primarily on personal contacts, is associated with rent discounts. Single person households, households with heads in the 20's, and households with eligible incomes also tend to be located in discount units.

Courant's interest in search was stimulated by his interest in explaining the impacts of discrimination. He argues that in a competitive housing market, racial discrimination should lead to a differential in housing prices such that whites who prefer segregated neighborhoods would pay a premium for living in such areas. In fact, however, not only are residential areas segregated, but blacks also often pay more for the same quality of housing than whites. Such a situation may be caused by perfect collusion among all whites against living and dealing with blacks, but this is not likely to occur. Courant argues that this paradox can be resolved by explicitly considering the cost of residential search.

In his model of residential search, buyers face a distribution of housing units with given characteristics (Courant, 1978). Presumably, buyers are assumed to maximize their utility by continuing the search until the expected benefit from further search is less than the expected cost of search. In this model, the existence of some sellers in a neighborhood who are averse to dealing with blacks will lower the expected benefit, hence, increase the expected cost of searching in that neighborhood. Most of the results are only implicitly related to questions of search. He does note that the maximum price differential is an increasing function of the proportion of white sellers unwilling to deal with blacks, and that if the price differential in a neighborhood is less than the maximum, blacks will not undertake any search in that neighborhood. In a study based on interviews of recent home-buyers in New Jersey, Lake (1980) also found that black housing search was less efficient and more costly than white; black households spent significantly more time in search, although considering fewer units in fewer neighborhoods.

Another theoretical study of real estate broker behavior discussses (1) the role of economic incentives for "steering" and other forms of racial discrimination by brokers, (2) the effect of multiple listing services on economic incentives and racial discrimination, and (3) some policy implications for reducing discrimination in broker behavior (Yinger, 1978). The framework for this analysis is a search model of broker behavior. A broker is faced with given probabilities of buyers and sellers deciding to use his services. It is assumed that the broker will attempt to maximize his income by individually adjusting the amount of search effort conducted on behalf of the client buyers and sellers and collectively (i.e., as a market) determining the optimal housing prices and commission rates. The paper draws the expected conclusions that:

(1) Brokers need to protect their reputation in order to attract future buyers and sellers, hence, they have an incentive not to introduce blacks to white neighborhoods.
(2) Multiple listing services can reduce the personal risk incurred by any one broker of damaging his reputation through irregular conduct, such as introducing blacks to listings in white neighborhoods.
(3) The government must conduct a more rigorous anti-discrimination effort in order to counteract the economic incentives of brokers to discriminate against blacks.

The paper by Smith and Mertz (1980) is concerned with both search and information. It is a bridge between the papers which study search per se and the analysis of information as a component of search (Clark and Smith, 1979; Maclennan, 1979). They use a basic model of housing search (Smith, Clark, Huff, and Shapiro, 1979) but allow the decision-maker's beliefs to be revised as market search occurs. The research shows that if an agent arranges vacancies in a non-random manner, he can significantly affect the search time and the price and quality of the vacancy that the client purchases (Smith and Mertz, 1980: 155).

## Information and Search

Empirical studies of the impact of information use on search behavior in the housing market have largely focused on the nature of information collection and information use. In most cases, the studies provide measures of the relative and absolute use of each of a number of information channels, including such information sources as real estate agents, newspapers, friends, co-workers, signs, and walking or riding around. A review of this material has indicated that there is considerable variation of the results between studies and there is almost no information on the sequence in which information sources are used and the interrelationship between channels (Clark and Smith, 1979). A recent study by Talarchek (1982) has gone some way towards addressing the issue of sequential use of information sources over time, but in general, there is little available data from the empirical studies about the way in which information channels are used over time. Those studies which have examined the temporal sequence have suggested that it involves newspapers, agents, friends, and relatives, in that order.

Attempts to model the household's search behavior in terms of the use of information channels and the searcher's cognitive modelling of channels have been discussed by Bettman et al. (1978) and Clark and Smith (1979) and Smith and Clark (1980). Bettman investigated the methods that individuals used to obtain information. He found that realtors and escrow agents dominated the information sources, although he also noted that consumers believed considerable information is unavailable to them. Amongst the identified sources the sample perceived as important (had they been available) were the fair value of house, the appreciation rate, the reputation of the real estate agent and a variety of neighborhood characteristics. In contrast with the empirical approach of Bettman et al., Clark and Smith attempted to design a housing market information flow system and a model of individual decision-making with respect to the acquisition and use of information. They were especially concerned to evaluate the effect of costs on the kinds of decisions that households made in terms of information channel selection. They were able to show that it was possible to model and simulate the use of multiple and interrelated channels of information flow in the housing market and that variations in cost are critical in determining the temporal and spatial sequences of search behavior. Even though the model is an obvious simplification of the market process, it provided considerable insight into the sequential structure and efficiency of search. For example, it could be shown that efficiency of the spatial search was particularly dependent on the quality of the realtor and the process of search with that agent.

At the present time, the work on search and information has been restricted to either the descriptive enumeration of sources of information largely drawn from survey analyses or the simulation studies illustrated in the work by Smith and Clark (1980). The role of information is currently of considerable interest in economics where optimality and equilibrium questions related to information have been examined by Riley (1979). The next step is to take the models that have been developed and tested by simulation data and to examine them in a market context. Until that is done, we cannot fully specify the way in which information sequentially modifies the search process.

## Stopping Rule Models

One of the earliest stopping rule models was developed by MacQueen and Miller (1960), who refer to it as the house-hunting problem. They assume that examining any alternative involves a fixed cost c, and that each alternative can be assigned a value v. The net payoff to the searcher who examines n alternatives is

$$\max (v_1, v_2, \ldots, v_n) - c\,n. \tag{2}$$

They show that the best strategy for the decision-maker is to establish a criterion value, $v^*$, and to accept the first alternative examined whose values exceeds $v^*$. If $f(v)$ is the distribution of the values of the alternatives, $v^*$ can be obtained from the equation

$$\int_{v^*}^{\infty} (v - v^*)\, f(v)\, dv = c. \tag{3}$$

Breiman (1964) has shown that linear programming can be applied to the solution of this stopping rule problem, and Friesz and Hall (1978) have discussed this connection in the context of residential mobility.

Phipps (1978) adopted this model in an experiment on students' apartment selection. His subjects examined up to fifty multi-attribute descriptions of apartments one by one, and had to try to select the best apartment they could. There was a non-zero probability that any apartment would no longer be available if it was not selected immediately on inspection. Some subjects recalled alternatives already examined, but most did not do so. Phipps found that only a few subjects followed the optimal strategy, perhaps because the costs of search were not included in the model.

A related model has been developed by Weibull (1978) for the search for housing. He assumes that alternatives arrive, or are located, at random time points with a probability density function $\lambda(t)$. Both $\lambda(t)$ and the distribution of values of alternatives are assumed to be known to the searcher. The criterion for stopping search, $v^*(t)$, and the cost of search, $c(t)$, are assumed to be functions of time. If the derivative of $c(t)$ is denoted $c'(t)$, the function $p(t) = -c'(t) / c(t)$ can be considered as a 'marginal impatience function'. If marginal impatience increases with time, the stopping criterion $v^*$ will be reduced.

In an attempt to overcome the difficulties of developing normative models of search, Meyer (1980) has suggested a less formalized model of decision-making under uncertainty. He justifies his model with the argument that behavior is not normative (Slovic, Fischhoff, and Lichtenstein, 1977) and that the normative models are difficult to test because of calibration difficulties. In a preliminary outline of a non-normative model, he makes a series of simple hypotheses about search behavior and specifically investigates how individuals form their utilities and how they gain their information. The model that Meyer develops is a stopping rule model with a different basis for the calculation of the utilities. While the cut-off level is similar to that used in models of optimal stopping, "the primary distinguishing feature of its use in its present context, is that here it is an empirical parameter unique to the individual and cannot be determined a priori as in the normative model" (Meyer, 1980: 24). In a preliminary test of this model in a situation of hypothetical apartment searches by undergraduate students, Meyer concluded that preferences change as the distribution of the utilities of opportunities is learned and that stopping is based on a process of making inferences about the distribution, the time available for search, and the quality of the alternative (apartment unit) that is being viewed. The emphasis on information processing and gathering is a strength of this model. Meyer reiterates the importance of further studies of modes of information gathering in search models.

Stopping rule models with hierarchical search have been developed for job search (Salop, 1973) in which the searcher has to choose a firm to sample before a job offer can be located. In this situation, he has to choose the order in which to sample the firms, based on the expected distribution of wages paid by each firm, and the probability that each firm will make a job offer. As might be anticipated, the expected gain from systematic search of this type is greater than the expected gain when job offers are sampled randomly.

A hierarchical search model was also developed by David (1974) for a somewhat different decision. The focus is still on job search but the searcher here must decide between sampling different local labor markets. If migration to a local labor market is regarded as a prerequisite for search to be carried on there, the cost of sampling that labor market is the cost of migration. The decision-maker knows the probability distribution of wage offers in each labor market, and chooses to migrate to one labor market. He then collects n job offers and selects the highest, recall being allowed. Miron (1978) provides a critique of David's work, including the suggestion that the job seeker in his new labor market should adopt the strategy of the house-hunting problem, and stop searching with the first offer exceeding a criterion value. He also argues that the job seeker may undertake a return trip to a labor market to look for work, rather than commit himself immediately to migration.

A different approach to hierarchical search has been discussed by Smith (1978). He argues that the decision-maker can reduce uncertainty by collecting information about the mean utility of alternatives in each of several areas, and that his strategy of spatial diversification in information collection can be compared to the financial problem of selecting a portfolio of investments.

One of the most ambitious applications of stopping rule models to housing market search is that of Smith, Clark, Huff and Shapiro (1979). They develop a hierarchical search model which is based on the searcher's subjective probability distributions for the value of dwelling units in each of a set of neighborhoods. At each stage in the search process, the searcher must decide whether to accept the best alternative available, with value $v_B$, or to continue search; in the latter case, he must select a neighborhood in which to search. Search will occur if the criterion value, $v^{*i}$, is positive for any neighborhood i, where

$$v^{*i} = E_t^i (v_B) - v_B. \tag{4}$$

$E_t^i (v_B)$ is the expected value of search at time t in neighborhood i, and is a function of the subjective probability distribution of the value of alternatives in neighborhood i and of the probability of losing the best alternative already located, as well as of $v_B$. Search is assumed to occur in that neighborhood where $v^{*i}$ is highest.

At this point, some comparisons can be drawn between the Smith et al. approach (which may be termed a hierarchical search model) and the disequilibrium models discussed earlier. They share the common assumption that households search and move in order to maximize their utility. However, the two models are directed toward different aspects of residential choice. The hierarchical search model is a model of search strategy, while the disequilibrium model is a model of housing demand and, by implication, of household mobility. The two models differ significantly in the following areas:

(1) The household's knowledge of the housing market. The disequilibrium model implicitly assumes that all households have a perfect knowledge of the housing market, while the hierarchical search model assumes only that households can estimate the distribution of prices and characteristics in each neighborhood.

(2) Search and moving costs. The disequilibrium model assumes fixed search and moving costs for all households of given characteristics, transforming search and moving costs into a transaction cost. The hierarchical search model includes a variable search cost, which acts as one of the constraints on search.

(3) The relationship between housing price and characteristics. The disequilibrium model typically assumes that housing price and characteristics are perfectly correlated, while

the hierarchical search model assumes only a joint distribution of prices and characteristics.

(4) Household demand for housing services. Using traditional economic analysis, the disequilibrium model derives demand for housing as a function of income, prices, and household characteristics. In the hierarchical search model, the unit price of housing services is not uniform, and demand for housing is less clearly defined.

### Spatial Aspects of Search

Following work by Barrett (1973) and Schneider (1975), Huff (1981) has specifically considered the spatial patterns of dwellings visited in the course of residential search. He suggests that the observed patterns will be related not only to the distribution of available vacancies but also to spatial biases in the search process, and he relates the location of vacancies visited to spatial models of search. He distinguishes between two approaches to spatial search patterns, one based on the probability of searching within each of several subareas of a city, and one based on the idea of anchor points. The area based model is discussed in this volume.

The anchor point model assumes that search is organised around some important place, such as a household's present residence, the work place or the real estate agent's office. Vacancies close to the anchor point are assumed to have a greater chance of being visited than those further away. Huff also discusses the shape of the distribution of vacancies around these anchor points, especially what he terms circular, teardrop, bimodal, and elliptical fields (Huff, 1981). He compares simulated patterns of vacancies (from the models of search) with observed patterns of search in the San Fernando Valley, Los Angeles.

## RESIDENTIAL SEARCH AND PUBLIC POLICY

The literature on housing search has been reviewed at length not only for its own sake but also to provide a background far some comments on the policy implications of search. Thus far policy interest in search processes has focused on issues of cost and discrimination. Both of these factors influence the degree of success in search. If individuals can find suitable accommodation in a quick and low cost search process, then there is no need for intervention in the process. On the other hand, if households search inefficiently for long periods of time and do not find suitable housing, they may be unable to move, or they may move to housing which does not fully meet their needs, and some policy intervention might be desirable.

In discussing policy issues with respect to search it is useful to subdivide the decision process. It can be divided into the decision to move, the process of search, and the choice of the best alternative. It can be assumed that the decision to look for a house is based on dissatisfaction with the present unit and on some prior knowledge of the housing market. Public policy probably has little direct influence on this first decision. However, periods of rapidly rising interest rates may stimulate people to look for housing before they might otherwise do so, or low vacancy rates may discourage households from searching. As in the case of the housing gap program of the Housing Allowance Demand Experiment, additional funds could encourage people to seek new housing.

Search can be regarded as an examination of alternative neighborhoods and houses in an effort to find the best available unit. It is assumed that a household continues to search until the expected benefit from further search is less than the expected cost of search (Smith et al., 1979). The expected costs include opportunity costs of search time, the probability of losing a previous alternative, and actual out-of-pocket expenses. Neither the cost nor the benefit of search can be computed with certainty, and so the search strategies depend on subjective evaluations of these factors.

The process of search can be characterized by the following factors:

(1) duration of search,
(2) number and type of information sources used,
(3) number of neighborhoods searched,
(4) number of houses examined, and
(5) radius of area searched.

In most economic studies these factors are grouped together to determine the general level of effort expended by a household in search. Search effort is considered equivalent to search costs. In general, it is expected that search effort will be positively related to household income. Higher income households should spend a greater amount of time and examine more alternatives than lower income families. Goodman (1978) reports a study by the National Association of Homebuilders which indicates that buyers of lower price houses (lower income buyers) examine fewer subdivisions than buyers of higher priced houses. However, McCarthy (1979) found no clear relationship between income and search effort, although higher income households tend to examine slightly more alternatives than the lower income households. The median number of search days spent by lower income households is, in fact, slightly higher than that spent by higher income households. The central policy issue is the differential role of constraints for high and low income households. Lower income renters are faced with more constraints on their search than higher income families, and the greater than expected search effort of lower income households may be a response to those constraints. There are two forms of constraints, those which affect the search process itself and those which may prevent the choice of the most appropriate alternative. In terms of constraint on the search process itself, lack of knowledge, lack of access or transportation, and lack of child care may be important. In terms of the final choice we can identify price constraints and discrimination.

What are the specific implications for public policy? All urban policy decisions have some effect on residential mobility. In particular, public programs which influence the quantity and distribution of housing have far reaching effects on the movement of households. Certainly, the suburbanization of the postwar era is a direct testimony to this statement. We can identify two possible policy objectives however, which are directly related to search models. The first objective might be to reduce the cost of residential search, especially for low income and minority households. A second objective could be to direct the search patterns of households, and hence their movements, to specific areas of the city. Both of these will be discussed briefly.

If the cost of search for low income and minority households is increased by constraints on the search process itself, then we can identify policy measures which might be directed towards reducing constraints and thereby improving search. Among these possibilities are the provision of active assistance (information, for example) in locating available vacancies, the provision of free transportation for households engaged in housing search, the provision of child care, and finally, the reduction of housing discrimination. Many measures designed to reduce constraints have already been tried in one form or another. In addition, most forms of housing discrimination, such as those based on sex, race, and nationality, are illegal.

Recent efforts to remove search constraints, however, have not had any discernable impact on moving behavior. Weinberg et al. (1977:112) could detect no measurable impact for the provision of assistance in the Housing Allowance Demand Experiments on the success of household search. Similarly, free transportation and child care services were ineffective in reducing search costs. The most surprising result was that the Housing Allowance Experiment had no significant impact on the household probability of undertaking search. There are, of course, very good reasons for this result and they have already been reported. On the one hand, the very existence of high transaction costs and the indivisibility of housing services may induce households to tolerate dissatisfaction or disequilibrium until some threshold is reached, and on the other, households base their housing expenditure on expected or permanent income and not on temporary additions to income. The enforcement of anti-discrimination legislation and the development of legislation to overcome discrimination based on age, source of income, and presence of children may be the most powerful force in affecting search and search costs.

Another goal of public policy seems to be to influence the choice of neighborhoods into which households migrate; for example, by attempting to stimulate the redevelopment of the central city by encouraging households to move there. The policy, to the extent that it exists, has been characterized as "public policy for places" (Goodman, 1978). It is an attempt to influence and upgrade neighborhoods rather than to influence the movement of people themselves. The evidence of previous studies of residential search and the results of a recent simulation study (Clark and Smith, 1979) indicate that households tend to confine their search to a limited area and that prior knowledge and personal contact with friends and relatives are among the most important factors in the choice of the area (Speare, Goldstein, and Frey, 1975:237).

These limited results suggest that public action could be applied to extend the household's area of familiarity. Since low income and minority households rely more on personal contact for sources of information, they may not be considering the full range of possible vacancies in parts of the metropolitan area not well known to them. Better presentation of information on these vacancies and the areas in which they are located could influence the search behavior of these households. This could be accomplished without major government intervention, independently of housing assistance or construction programs. Moreover, changing the format and availability of information provision may, in the long run, have a bigger impact on search patterns and housing choices than more direct policy intervention.

## CONCLUDING REMARKS

Several common threads emerge from this review of the search literature and bear repeating in these concluding remarks. First, most search studies are cross-sectional. There have been only a few longitudinal analyses of the search process, and the process itself is still not well-specified. Also, the temporal analyses are quite restrictive in their assumptions. Second, in almost all the analyses of the search process (even in the most recent cases which are discussed in this paper) space is very broadly defined. In most of the work, there is no spatial dimension. The process of search is seen as sequential behavior and only recently has space been introduced in the analyses of search processes. Apart from some preliminary observations by Weibull (1978) and the recent work by Huff (1981), there is little attempt to deal with space in a real, as distinct from an abstract form. This may be partly a function of the fact that space is a very complex variable and difficult to introduce into a search model. There are sufficient difficulties in specifying the process temporally; the spatial complication has been skirted over by most people attempting to build models of search. Third, in most of the search studies, the emphasis has been on some form of utility maximization to search (that is, on consumer or demand approaches) or on institutional roles which bias search. But, in the work to date, the nature of institutional constraints have not been formally incorporated into models of search. This remains an area for considerable research. Fourth, the work on policy outcomes of search and search behaviors has focused largely on cost and discrimination. These analyses, as Moore makes explicit in a later chapter in the book, are only a partial approach to the myriad and complex approaches to search from a policy perspective. Finally, the research which deals specifically with housing market search has established a useful distinction amongst the decision to search, studies of the process of search (which are often quite empirical in their focus) and analyses of the choice of the best alternative. It is the latter which is often quite complex and mathematically sophisticated, especially the work on optimal stopping rules.

## REFERENCES


Adams, J. S. (1969) "Directional bias in intraurban migration", Economic Geography, 45, 302-323.

Anderson, N. (1968) "A simple model for information integration", in R. P. Abelson et al. (eds.), Theories of Cognitive Consistency: A Sourcebook. Chicago: Rand McNally.

----- (1974) "Information integration theory: a brief survey", in D. H. Krantz et al. (eds.), Contemporary Developments in Mathematical Psychology, 2, San Francisco: W. H. Freeman.

Barrett, F. (1973) Residential Search Behavior. Toronto, Canada: York University, Geographical Monograph Number 1.

Bettman, J. R. (1976) "Data collection and analysis approaches for studying consumer information processing", Advances in Consumer Research, 3, 342-348.

----- (1979) An Information Processing Theory of Consumer Choice. Reading, Massachusetts: Addison Wesley.

----- N. Capon, R. Lutz, G. Belch, and M. Burke (1978) "Affirmative disclosure in home purchasing", Los Angeles: University of California Los Angeles, Graduate School of Management, Occasional Paper Number 14.

Breiman, L. (1964) "Stopping-rule problems", in E. F. Beckenbach (ed.), Applied Combinatorial Mathematics. New York: Wiley.

Brickman, P. (1972) "Optional stoppping on ascending and descending series", Organizational Behavior and Human Performance, 7, 53-62.

Chinloy, P. T. (1980) "An empirical model of the market for resale homes", Journal of Urban Economics, 7, 279-292.

Clark, W. A. V. (1981) "On modelling search behavior", in D. Griffith and R. MacKinnon (eds.), Dynamic Spatial Models. New York: Sijthoff and Noordhoff.

----- and T. R. Smith (1979) "Modelling information use in a spatial context", Annals of the Association of American Geographers, 69, 575-588.

----- and T. R. Smith (1982) "Housing market search behavior and expected utility theory II: the processes of search", Environment and Planning A, (forthcoming).

Courant, P. N. (1978) "Racial prejudice in a search model of the urban housing market", Journal of Urban Economics, 5, 329-345.

Cronin, F. J. (1979a) "An economic analysis of intraurban search and mobility using alternative benefit measures", Washington, D. C.: The Urban Institute.
----- (1979b) "Low income households' search for housing: preliminary findings on racial differences", Washington, D. C.: The Urban Institute.
David, P. A. (1974) "Fortune, risk, and the microeconomics of migration", in P. A. David and M. W. Reder (eds.), Nations and Households in Economic Growth. New York: Academic Press.
Edwards, W. (1954) "The theory of decision making", Psychological Bulletin, 15, 380-417.
Flowerdew, R. (1976) "Search strategies and stopping rules in residential mobility", Institute for British Geographers Transactions, 1, 47-57.
----- (1978) "The role of time in models of residential choice", in T. Carlstein, D. N. Parkes, and N. J. Thrift (eds.), Human Activity and Time Geography. London: Edward Arnold.
Friesz, T. L. and P. D. Hall (1978) "Residential choice, optimal stopping and linear programming", Northeast Regional Science Review, 8, 65-83.
Goodman, J. L., Jr. (1976) "Housing consumption disequilibrium and local residential mobility", Environment and Planning A, 8, 855-874.
----- (1978) Urban Residential Mobility: Places, People and Policy. Washington, D. C.: The Urban Institute.
Gould, P. (1966) Space Searching Procedures in Geography and the Social Sciences. Honolulu, Hawaii: Social Science Research Institute.
Granovetter, M. S. (1974) Getting a Job: A Study of Contacts and Careers. Cambridge, Massachusetts: Harvard University Press.
Hanushek, E. A. and J. M. Quigley (1978) "An explicit model of intra-metropolitan mobility", Land Economics, 54, 411-429.
Hartshorn, T. A. (1980) Interpreting the City. New York: Wiley.
Hay, A. M. and R. J. Johnston (1979) "Search and choice of shopping centre: two models of variability in destination selection", Environment and Planning A, 11, 791-804.
Herbert, D. T. (1973) "The residential mobility process: some empirical observations", Area, 5, 44-48.
Huff, J. O. (1981) "Patterns of residential search", Los Angeles, California: Paper presented at the Annual Meeting of the Association of American Geographers.
Ioannides, Y. M. (1975) "Market allocation through search: equilibrium adjustment and price dispersion", Journal of Economic Theory, 11, 247-249.
Kahan, J. I., A. Rapoport, and L. V. Jones (1967) "Decision making in a sequential search task", Perception and Psychophysics, 2, 374-376.
Kahn, L. M. (1978) "The returns to job search: a test of two models", Review of Economics and Statistics, 60, 496-503.

Kleiter, G. D. (1975) "Dynamic decision behavior: comments on Rapoport's paper", in D. Wendt and C. A. J. Vlek (eds.), Utility, Probability and Human Decision Making. Dordrecht, The Netherlands: Reidel.

Kohn, M. G. and S. Shavell (1974) "The theory of search", Journal of Economic Theory, 9, 93-123.

Lake, R. W. (1980) "Housing search experiences of black and white suburban homebuyers", in G. Sternlieb et al. (eds.), America's Housing: Prospects and Problems. New Brunswick, N. J.: Rutgers University, Center for Urban Policy Research.

Levine, M. J., M. G. Samet and R. E. Brahlek (1975) "Information seeking with limitations on available information resources", Human Factors, 17, 502-513.

Lippman, S. A. and J. J. McCall (1976) "The economics of job search: a survey", Economic Inquiry, 14, 155-189 and 347-368.

----- and J. J. McCall (1979) Studies in the Economics of Search. Amsterdam: North-Holland.

Maclennan, D (1979) "Information networks in a local housing market", Scottish Journal of Political Economy, 26, 73-88.

MacQueen, J. and R. G. Miller Jr. (1960) "Optimal persistence policies", Operations Research, 8, 362-380.

McCall, J. J. (1970) "Economics of information and job search", Quarterly Journal of Economics, 84, 113-126.

McCarthy, K. F. (1979) "Housing search and mobility", Santa Monica, California: Rand Corporation Report R-2451-HUD.

McCracken, K. (1975) "Household awareness spaces and intraurban migration search behavior", The Professional Geographer, 27, 166-170.

Melnik, A. and D. H. Saks (1977) "Information and adaptive job search behavior: an empirical analysis", in O. C. Ashenfelter and W. E. Oates (eds.), Essays in Labor Market Analysis. New York: Wiley.

Meyer, R. (1980) "A descriptive model of constrained residential search", Geographical Analysis, 12, 21-32.

Miron, J. R. (1978) "Job-search perspectives on migration behaviour", Environment and Planning A, 10, 519-535.

Moore, E. G. (1970) "Some spatial properties of urban contact fields", Geographical Analysis, 2, 376-386.

----- and L. A. Brown (1970) "Urban acquaintance fields: an evaluation of spatial models", Environment and Planning A, 2, 443-454.

Palm, R. (1976a) "Real estate agents and geographical information", The Geographical Review, 66, 266-280.

----- (1976b) "The role of real estate agents as information mediators in two American cities", Geografiska Annaler, 5B, 28-41.

Phipps, A. G. (1978) "Space searching behavior: the case of apartment selection", Iowa City, Iowa: University of Iowa, Unpublished PhD Dissertation.

Pissarides, C. A. (1976) Labour Market Adjustment: Microeconomic Foundations of Short-Run Neoclassical and Keynesian Dynamics. Cambridge, England: Cambridge University Press.

Rapoport, A. (1975) "Research paradigms for studying dynamic decision behavior", in D. Wendt and C. A. J. Vlek (eds.), Utility, Probability and Human Decision Making. Dordrecht, The Netherlands: Reidel.

----- R. W. Lissitz, and H. A. McAllister (1972) "Search behavior with and without optional stopping", Organizational Behavior and Human Performance, 7, 1-17.

Riley, J. (1979) "Informational equilibrium", Econometrica, 47, 331-359.

Rossi, P. (1955) Why Families Move: A Study in the Social Psychology of Urban Residential Mobility. New York: The Free Press.

Rothschild, M. (1973) "Models of market organization with imperfect information: a survey", Journal of Political Economy, 81, 1283-1308.

Salop, S. C. (1973) "Systematic job search and unemployment", Review of Economic Studies, 40, 191-201.

Schiller, B. R. (1975) "Job search media: utilization and effectiveness", Quarterly Review of Economics and Business, 15, 55-63.

Schneider, C. H. P. (1975) "Models of space searching in urban areas", Geographical Analysis, 7, 173-185.

Silk, J. (1971) "Search behavior: general characterization and review of the literature in the behavioral sciences", Reading, England: University of Reading, Department of Geography, Geographical Paper, Number 7.

Slovic, P., B. Fischhoff and S. Lichtenstein (1977) "Behavioral decision theory", Annual Review of Psychology, 28, 1-39.

----- and S. Lichtenstein (1971) "Comparison of Bayesian and regression approaches to the study of information processing in judgement", Organizational Behavior and Human Performance, 6, 649-744.

Smith, T. R. (1978) "Uncertainty, diversification, and mental maps in spatial choice problems", Geographical Analysis, 10, 120-141.

----- W. A. V. Clark, J. O. Huff, and P. Shapiro (1979) "A decision-making and search model for intraurban migration", Geographical Analysis, 11, 1-22.

----- and W. A. V. Clark (1980) "Housing market search: information constraints and efficiency", in W. A. V. Clark and E. G. Moore (eds.), Residential Mobility and Public Policy. Beverly Hills, California: Sage Publications.

----- and W. A. V. Clark (1982) "Housing market search behavior and expected utility theory I: measuring preferences for housing", Environment and Planning, (forthcoming).

----- and Mertz, F. (1980) "An analysis of the effects of information revision on the outcome of housing market search, with special reference to the influence of realty agents", Environment and Planning A, 12, 155-174.

Speare, A., S. Goldstein and W. Frey (1975) Residential Mobility, Migration and Metropolitan Change. Cambridge, Massachusetts: Ballinger.

Stigler, G. J. (1961) "The economics of information", Journal of Political Economy, 69, 213-225.

Talarchek, G. (1982) "A model of residential search and selection", Urban Geography, (forthcoming).

Telser, L. G. (1973) "Searching for the lowest price", American Economic Review, 63, 40-49.

Weibull, J. W. (1978) "A search model for microeconomic analysis -- with spatial applications", in A. Karlqvist, L. Lindqvist, F. Snickirs, and J. W. Weibull (eds.), Spatial Interaction Theory and Planning Models. Amsterdam: North-Holland.

Weinberg, D., R. Atkinson, A. Vidal, J. Wallace and G. Weisbrod (1977) Housing Allowance Demand Experiment, Locational Choice, Part I, Search and Mobility. Cambridge, Massachusetts: Abt Associates Inc.

----- J. Friedman and S. K. Mayo (1979) "A disequilibrium model of housing search and residential mobility", Cambridge, Massachusetts: Abt Associates Inc.

Wolpert, J. (1965) "Behavioral aspects of the decision to migrate", Papers of the Regional Science Association, 15, 159-169.

Yinger, J. (1978) "Economic incentives, institutions, and racial discrimination: the case of real estate brokers", Cambridge, Massachusetts: Harvard University, Department of City and Regional Planning, Discussion Paper D78-4.

# 2

# AN ANALYTICAL MODEL OF HOUSING SEARCH AND MOBILITY

**Kevin McCarthy**

Despite its position in the residential mobility process, housing search as a determinant of moving behavior has been given only brief attention in literature on the subject. Although limitations of data have contributed to this oversight, in large part it reflects the narrowness of the traditional analytic approaches, which typically focus on moving per se and thus simplify what is in fact a complex and multifaceted mobility process. As a result, even those studies which acknowledge the importance of the search process have failed to focus on it in a satisfactory way. Such studies divide into two types: formal models of the decision to move (Speare et al., 1975; Hanushek and Quigley, 1978) that recognize housing search as a transaction cost but rarely examine search behavior; and descriptive studies of search activity (Barresi, 1968; Hempel, 1969a and 1969b; Barrett, 1973) that lack a satisfactory theoretical structure for assessing how search affects mobility. Consequently, too little is understood about how households' moving decisions are shaped by the perceived benefits and costs of moving, how households' uncertainty about those benefits and costs influences their decisions whether and how to undertake an active search, or how various search costs affect moving behavior. (In the very recent past papers by Smith et al., (1979) and Meyer (1980) have gone some way towards filling this gap.)

The substantive importance of the search process is suggested by several explicit mobility models that have appeared in the literature (Speare et al., 1975; Hanushek and Quigley, 1978; Huff and Clark, 1978; Brummell, 1979 and Weinberg et al., 1981). Although differing on specifics, these models share certain common features: (1) a behavioral approach, focusing on the separate influences on the decision-making process; (2) recognition (implicitly) of mobility as a mechanism of housing adjustment, with the inclusion of consumption measures in the decision-making process; and (3) a focus on the decision to move per se, typically ignoring the type of consumption adjustment that moving produces. By focusing on the behavioral aspects of mobility and its role in the housing adjustment process, these models emphasize the consumption rather than the spatial aspects of moving. As such, they represent a significant advance over prior models that regarded local residential mobility and


DOI: 10.4324/9781003182085-3

migration, both of which entail spatial relocation, as fundamentally alike.

Despite this valuable insight, those models rarely examine in any detail the "consumer behavior" that mobility entails, that is, how a household shops for a new residence. This is a significant omission, because moving is a complex behavior entailing a series of choices rather than a single decision or behavior. Those choices, which may not all be present in every case, include the decisions to consider moving, to undertake an active search, and whether and where to move (Rossi, 1955; Brown and Moore, 1970). Because the search stage intervenes between the decision to consider moving and the actual move, the characteristics of the search are likely to play an important role in determining whether households are able to make their desired adjustments when they move. Moreover, insofar as the determinants of the separate stages differ, an understanding of the search process is essential to a complete analysis of moving behavior (Wolpert, 1965).

Although no comprehensive analysis of transaction costs currently exists, several recent empirical studies (Newman and Duncan, 1979; McCarthy, 1979 and 1980; Weinberg et al., 1981) and at least one theoretical paper (Smith et. al., 1979) have suggested that the transaction costs of moving--that is, the time, effort, and monetary costs involved in locating and moving to a new residence are substantial and can significantly affect moving behavior. Without a systematic analysis, however, it is impossible to determine which types of transaction costs are most significant, how they affect a household's moving behavior, or what might be done to overcome those effects.

This paper describes a conceptual framework and an explicit model designed to facilitate a systematic analysis of the housing search process and its effects on moving behavior. This model focuses on the following issues: why do households search, what procedures do they use, do the transaction costs of moving affect mobility and if so, which costs are most important, what types of households are most sensitive to specific costs and how do those costs affect different aspects of the consumption adjustment process. Answers to these questions will not only provide important insights into the mobility processes but they will also provide useful information for policymakers concerned with expanding the residential choices of policy-relevant households, e.g. those who are elderly or minorities, or have children or low incomes.

The next section, introducing the search model, begins with a discussion of the concepts underlying the model. It then compares this framework with others in the literature. It concludes with a presentation of the model and a discussion of its strengths and weaknesses. The final section, then presents evidence on how search procedures vary, how supply conditions affect search procedures, and how search procedures influence moving outcomes.

## THEORETICAL FRAMEWORK

Our underlying conceptual model of the mobility process shares certain characteristics with recent behavioral models of mobility, but extends them by emphasizing transaction costs and how they impinge on moving behavior. This section describes the model, beginning with a statement of its conceptual underpinnings and how they compare with those of other approaches. We then introduce the three-stage search model.

### Underlying Assumptions

Several assumptions about the residential mobility process underlie the search model used here. These assumptions relate to: (1) the household's efforts to adjust its consumption of housing, (2) its position vis-a-vis a hypothetical equilibrium between actual and desired housing circumstances, (3) the search costs a household is willing to absorb, and (4) the separate decisions which, in sequence, culminate in a move.

Our first assumption posits that most local moves are motivated by a household's desire to adjust its housing consumption. In this respect, residential mobility contrasts sharply with migration, since migrants are typically viewed as investors in their own human capital who move in anticipation of the employment and income benefits to be reaped at their new destinations (Da Vanzo, 1976; Greenwood, 1975). From our perspective the spatial aspects of residential mobility are incidental to its function as a mechanism for consumption adjustment: What the household is doing relates first and foremost to housing. Whether or not this adjustment entails moving depends on how the household perceives the advantages and disadvantages (benefits and costs) associated with moving. Households implicitly weigh those benefits and costs and move only when it seems advantageous to do so. Benefits here include the housing and neighborhood improvements that may be realized by moving, such as more space or a safer neighborhood. Costs include those required to find alternative housing and then to change residences. Specific search and relocation costs may consist of direct expenses (e.g. for transportation or moving household goods), opportunity costs (e.g. for the time and effort spent searching), and psychological costs (due to discrimination encountered during search or breaking neighborhood attachments upon moving).

Our second assumption posits the notion of equilibrium. The benefits of moving, and thus the likelihood that a household will contemplate a move, will depend partly on how distant the household is from (or close to) some hypothetical state of equilibrium between its desired and actual housing circumstances. Depending on that balance, households may seek to improve the fit between what they have and what they need, either by increasing or reducing their level of consumption. For example, a young couple expecting a child may need another bedroom, whereas an older couple whose children have left home may find they are consuming and paying for more than they need.

Third, we assume that because households never have perfect information with which to make their housing choices, they typically search out alternatives to gain a better assessment of the benefits and costs of moving. How they conduct this search occupies a central place in our conceptual framework. Specifically, we assume that the household embarks on a search without knowing how much searching will be necessary or even whether it will prove sufficiently fruitful to justify the effort. We further assume that difficulties experienced during the search, particularly discrimination, may force households to revise their original expectations, modify their moving goals, or even to terminate their search and postpone moving. The search costs that a household is willing to absorb will depend on the benefits it expects to receive and how long it expects to receive them.

Finally, this framework explicitly assumes that residential mobility typically entails a series of analytically separate decisions or behaviors, including the decision to consider moving, the decision to undertake an active search, and the decisions of whether and where to move. By explicitly recognizing that more than one decision is involved in the mobility process, this approach also acknowledges that there is more than one behavior to explain and that the determinants of each behavior need not be the same. This final point is especially important, because many analysts have restricted their focus to the single variable--whether households actually move. However, if the importance of a particular class of variables changes at different stages of the mobility process, a model using a single dependent variable will not be able to capture that change and may obscure the importance of that class of factors altogether.

## Comparison With Other Analytical Frameworks

In its general outline, this framework closely resembles those of other mobility models. For example, most models, acknowledging the reasons prior research has demonstrated households report for moving (Morgan, 1972; Bureau of Census, 1966), agree that residential mobility is primarily consumption related. Similarly, most models contain some notion of benefits and costs, assume that households will consider moving when they believe some other residence offers greater benefits than their current units, and will decide to move when the expected benefits of moving exceed its costs.

Of course, the terminology used to refer to benefits and costs, how they are measured, and the specific advantages and disadvantages of particular models often differ considerably (Quigley and Weinberg, 1977). Economists, for example, assert that the probability of moving is a function of the expected gain in utility which households receive by bringing their actual consumption into closer balance with their "equilibrium" level of consumption (moving benefits) and expected moving costs ( Abt, 1978; Hanushek and Quigley, 1978; Cronin, 1978). By assigning explicit dollar values to both benefits and costs, the economic models attempt to obtain a more objective measure of the net benefits of moving than models that rely on more subjective measures.

There are, however, some potentially serious problems with the way the economic models measure both benefits and costs. Potential moving benefits, for example, are typically estimated in terms of the difference in the volume of services households consume in their current residences and an estimated equilibrium volume of services. The greater this difference, whether positive or negative, the greater the presumed utility households can gain by moving. The equilibrium level of services is, in turn, estimated in terms of total housing expenditures by assuming that expenditures, when adjusted for price discounts, and so on, accurately capture differences in the volume of services consumed. However, the utility a household derives from housing is determined not simply by the total volume of services consumed but also (and perhaps more importantly) by consumption of specific attributes of the housing bundle. Since housing is a multi-dimensional good and different combinations of attributes can command the same price, measures of volume alone can never measure the household's expected utility gain. This problem may contribute to the fact that in most economic models, measures of costs typically perform better than measures of benefits (Cronin, 1978; Weinberg et al., 1979 and 1981).

There are also problems with the economic model's attempt to measure moving costs. Although such costs as direct relocation expenses, length of stay discounts, and closing costs either are already or can easily be measured in dollars, the value of search time and effort (which are measured in terms of their opportunity costs), and the psychological costs of discrimination are much harder to evaluate in dollar terms. Finally, by collapsing the cost measures into a single value, the economic model often overlooks the relationships between these costs (e.g., discrimination and search effort), and thus the opportunity to determine the impact of individual cost items on moving behavior.

Geographers and sociologists, on the other hand, although utilizing similar concepts, define and measure them in different ways. Geographers, for example, refer to residential stress and resistance to moving rather than benefits and costs. Residential stress refers to the pressure to move arising from a household's dissatisfaction with its residence. Stress is measured in terms of the household's evaluation of various attributes of its unit and location, or what is referred to as the household's experienced place utility (Clark and Cadwallader, 1973). The difference between experienced place utility and aspiration place utility (the amount of stress relieved by moving) defines the benefits of the move (Brummell, 1979). Mobility resistance (moving costs) is less well defined, but unlike costs in the economic model, includes both monetary and nonmonetary elements; for example, both direct relocation expenditures and the emotional costs of breaking ties to prior residences (see Wolpert, 1965; Brown and Moore, 1970; Huff and Clark, 1978; Smith et al., 1979).

Sociologists refer to residential satisfaction rather than utility or residential stress, and assume that the benefits to be gained are reflected in the increased satisfaction that results from moving. Measures of a household's expected benefits are obtained by asking respondents how satisfied (or dissatisfied) they are with the current residence and whether they plan to move. Like geograpahers, sociologists have not clearly delineated the costs of moving but agree they are important and include both monetary and nonmonetary factors. Variables that identify those characteristics of households that might inhibit their mobility are often used in sociological and geographic models to capture the effects of moving costs (Rossi, 1955; Speare, 1974; Speare et. al., 1975).

Although neither the geographic nor sociological models of mobility contain a direct equivalent of the economist's notion of equilibrium, both assume that households implicitly weigh their level of residential stress or dissatisfaction against some intuitively recognized threshhold or aspiration level in deciding whether to move. When the level of either stress or dissatisfaction exceeds the threshhold, the probability of moving increases. Since the household's threshold or aspiration level, like the economist's notion of equilibrium consumption, is assumed to be determined by household characteristics, the household's level of stress or dissatisfaction relative to its threshhold operates in essentially the same way as the economist's notion of disequilibrium.

By incorporating multiple indicators of housing and neighborhood characteristics, the geographic and sociological models are better equipped than the economic model to identify which aspects of a household's current housing circumstances have the strongest effect on its overall mobility. Moreover, the subjective nature of some housing circumstance variables used in these models may be a more appropriate determinant of mobility (or at least of the decision to search) than the single "objective" benefit measure, because before a household actually searches, the best indicator of whether it could benefit from moving is its current housing circumstances. Indeed, two studies that included both objective and subjective measures of housing circumstances found the subjective variables to be better predictors of the decision to search (Cronin, 1978; Weinberg and Atkinson, 1979). However, the treatment of moving costs in these models has several weaknesses. First, it obscures the benefit/cost tradeoffs inherent in the adjustment process by attempting to capture those tradeoffs indirectly. Second, by using household characteristic variables to capture the effects of moving costs, these models potentially confound the effects of household characteristics on moving costs with their effect on the propensity to experience changes in circumstances that could also trigger mobility. Finally, by excluding direct measures of moving costs, this approach offers no policy leverage for changing mobility behavior.

An analytical model of search

Each of the approaches discussed here recognizes that a household's decision to move is only the final stage of a complex process. However, only some of those models incorporate all of those stages and few, if any, analyze how the process itself affects the housing adjustments movers make. As a result those approaches run the risk of misspecifying the effects of a particular variable that influences only one stage of the process and miss the opportunity to identify how the process itself shapes the eventual outcome. Thus, it may be difficult with such models to explain why some households who are dissatisfied with their current housing do not search, why some households who search do not move, or how difficulties experienced during the mobility process, such as discrimination, affect its outcomes.

## A THREE-STAGE SEARCH MODEL

While incorporating elements from other mobility models, the approach used here emphasizes transaction costs and their effects by employing what might be called a three-stage search model. This model separates moving into three distinct stages: first, the decision to engage in an active search; second, the search itself; third, the outcome of the search. The outcome stage of the model is actually divided into two phases: the first focusing on whether a move occurs; the second on the nature of the housing adjustment that movers make. The model is designed to explain which households search, what procedures they use, and how these procedures influence moving behavior. (By focusing on moving among searchers rather than among all households, this approach necessarily loses total closure on mobility, since not all households search before moving. This exclusion is purposeful. This research focuses on how the search process affects moving behavior. It would clearly be inappropriate to include "windfall" movers--those who did not search.)

The first stage estimates the probability that a household will conduct an active search as a function of its current housing circumstances, $h_i$, its demographic and economic characteristics, $g_i$, and its prior market knowledge, $k_i$:

$$P(S) = f(h_i, g_i, k_i) \qquad (1)$$

The benefits of moving are introduced into this model through the housing circumstance variables ($h_i$), which include both objective condition and subjective rating measures. Underlying this approach to measuring expected moving benefits is the assumption that before a household actually searches, the best indicator of whether it can benefit from moving is its own evaluation of its current housing situation. Moreover, multiple measures avoid the problems of a single predictor of benefits and enable us to determine whether selected housing problems, e.g. high costs, crowding, unsafe neighborhoods, provide a greater stimulus for moving than do others.

Before engaging in an active search, households are unlikely to have a firm estimate of their search and relocation costs and thus the net benefits of moving. However, we assume that households can and do form expectations about those costs before they embark on a search. Specifically, before deciding to search households must consider such factors as the probable time and effort required to find a suitable unit, the likelihood of encountering special search problems--whether due to their own circumstances, e.g. age, employment, child care, or by the actions of others, e.g. discrimination, as well as the likely costs of moving to a new unit. A household's expectations about these factors will be based on an assessment of its own circumstances and characteristics as measured by the $g_i$ variables.

Finally, we include measures of prior market knowledge ($k_i$) on the assumption that a household's need to conduct an active search will depend on its prior familiarity with the market. Thus, recent movers, having already become familiar with the alternatives available, may have less need to conduct an active search than those whose market information is out-of-date.

With respect to the second stage, the search procedures used, we assume that when households embark on a search, they are in effect gambling. They can only guess at how much searching will be necessary, and the outcome may not justify their efforts. These uncertainties lead households to adopt widely different strategies for deciding how much effort to expend and what information sources to use. At one extreme, the costs of searching might be minimized by abstaining from any activity at all--essentially doing nothing more than remaining alert to "windfall" discoveries picked up from information from friends or casual perusal of the market. At the other extreme, a household might maximize its chances of locating the best available alternative by continuous and thorough search--looking for months and considering dozens of alternatives. Most households fall between these two extremes, of course, or alter their search procedures as they become familiar with what the market has to offer.

Our model assumes that the search strategies households adopt will be influenced by the same three factors that determine whether a household undertakes an active search: (1) current housing circumstances ($h_i$), (2) demographic and economic characteristics ($g_i$), and (3) familiarity with the market ($k_i$). Unlike the decision to search, however, search strategies will also be shaped by (4) events occuring during the search ($d_i$).

$$\text{Search Procedures} = f(h_i, g_i, k_i, d_i) \qquad (2)$$

A household's evaluation of its current housing circumstances will influence its choice of strategy by shaping its expectations of the benefits it can expect from moving and thus, the search costs it can reasonably afford. Households whose current housing is generally satisfactory, and thus might expect only minimal benefits from moving, may be only "passively alert," but those that are in substantial disequilibrium can be expected to search actively.

Household characteristics will influence search procedures in several ways. First, they determine the type of unit sought, and, since appropriate strategies vary, the search procedures likely to be adopted. Prospective home buyers, for example, consult real estate agents more frequently than searchers looking for rental units. Second, they affect the costs of using alternative strategies. For example, some households, whether due to age or employment, may find it particularly costly to visit personally a large number of alternatives and rely instead on newspaper advertisements or referrals from friends, while others may find their best strategy is to rely on their own efforts. Third, although most households move to adjust their housing consumption, the frequency with which they are required to make those adjustments varies with their characteristics. Correspondingly, households who can expect to make frequent adjustments should be unwilling to expend the same effort to find a new residence as those whose circumstances are more stable. Finally, households whose characteristics (e.g. race, family status, etc.) increase the likelihood that they will encounter discrimination during their search may tailor their search procedures accordingly.

A household's previous familiarity with the market should also influence its search strategy. Most households, since they enter the market infrequently, are unfamiliar with the options available. They must first explore the market to establish criteria for choosing a new unit and then locate and rank alternatives (Silk, 1971). Some households, however, have recently searched for housing and their prior experience should reduce the effort they must expend to locate an acceptable unit.

Finally, a household's initial expectations as to its search and relocation costs may be modified by difficulties encountered after it begins searching. Such problems may cause a household to alter its initial strategy or even abandon its plans to move altogether. For the most part, these problems are of the type consumers generally face when they enter a market and can be attributed to such things as inadequate market knowledge, limited supply, etc. However, some households face special difficulties in their search because they are discriminated against in the market. Whether due to race, income, or family circumstances, discrimination increases a household's search costs by subjecting it to humiliation or hostility and forcing it to expend more effort to find a suitable residence (McCarthy, 1980).

The third stage of the model focuses on the outcomes of the search. Unlike traditional models that measure the outcome of the search in terms of whether or not households move, our model distinguishes between the decision to move per se and the nature of the adjustment households make when they move. Thus, we separate the final stage of the search model, the outcome of search into its two analytically distinct phases. There are two reasons for this approach. First, the determinants of the decision to move and the choice of a new unit are not identical. Second, search and relocation costs are unlikely to have comparable impacts on these two decisions.

Unlike the choice of a new unit, the decision to move is directly related to a household's current housing circumstances, since it is the lack of fit between the household's current consumption and its desired or equilibrium consumption that motivates the decision to move. The type of adjustment a household makes when it moves, on the other hand, bears no necessary relationship to its prior housing circumstances because it is the household's current housing needs that determine its desired consumption adjustment. Thus, it is not surprising that the explanations local movers give for deciding to move do not correspond at all closely with their reported reasons for choosing their new residences (Butler et al., 1969).

Although transaction costs can affect both the decision to move and they type of adjustment made, it is unlikely that these two effects will be comparable. Probably the most severe effect moving costs can have on mobility is to dissuade a household that has actively searched, from moving. Such households receive nothing in return for the time and energy invested in searching and may be forced to tolerate the conditions that first caused them to search. (Alternatively, households who stop searching without moving may decide to improve their current unit or change the household circumstances that prompted the decision to search.) Rather than causing searchers to forego moving, a more likely effect of moving costs is that they will affect the size and character of the adjustment made. For example, moving costs may persuade a household to select the first acceptable unit rather than continue to search for the best available unit. Alternatively, moving costs could force households to modify their consumption of the various components of the housing bundle; for example, a household experiencing discrimination when searching in a perferred neighborhood might decide to trade off neighborhood quality for a larger and a better quality unit in its original neighborhood.

Since our model treats the decision to move and the type of adjustment made as separate outcomes of the search process and since the determinants of these two outcomes are assumed to differ, two separate equations are used to model these outcomes. The first equation focuses on the decision to move and estimates the conditional probability that a household will move given that it conducted an active search. This probability is assumed to depend on current housing circumstances, $h_i$, the costs of searching, $c_i$, the search procedures used, $s_i$, and household characteristics, $g_i$,

$$P(M/S) = f(h_i, c_i, s_i, g_i) \tag{3}$$

The housing circumstance measures are once again assumed to identify the potential benefits of moving. The search-cost variables capture the costs of searching and any special problems including discrimination that households may have experienced during the search. The search-procedure variables identify the number and types of information sources used and the areal extent of the search. Finally, the household characteristic variables provide a necessary control for those factors.

The other outcome equation then estimates how relocation and search costs affect the type of adjustments households make when they move. Those adjustments are assumed to depend principally on the household characteristics, $g_i$, (including any changes in characteristics) that may have precipitated the move. However, a variety of movings costs, the $c_i$'s, including discrimination, may force households to modify the nature of the adjustment they make. Similarly, some search techniques and procedures may be more effective than others. Consequently, variables identifying those factors, the $s_i$, are also included in the model.

$$\text{Type of Adjustment} = f(c_i,\ s_i,\ g_i) \qquad (4)$$

No specific consumption change measure is included in equation 4, because several different dimensions of that adjustment can be examined. Those dimensions include objective measures of the difference in consumption between pre- and postmove units, $H_0$-$H_1$, and of the price households pay for their new housing. The objective measures of consumption change, including the total change in the volume of housing services consumed and the change in each of the three principal dimensions of the housing bundle (space, unit quality, and neighborhood), can be estimated using hedonic index equations, which estimate the price on the various attributes of the housing bundle based on their contribution to total rent, and the characteristics of old and new residences. Hedonic indexes can also be used to determine if renters pay a premium for their new units or get them for a bargain. Since the hedonic prices are equivalent to the average rent for any given housing attribute, differencing the actual rent from an estimated rent obtained by multiplying attributes by their predicted prices will indicate whether a household is paying more or less than average rent for a given unit, that is, whether they paid a premium or found a bargain. (This technique is demonstrated in McCarthy (1980).)

Each of these "objective" measures takes into consideration only the outcomes of moving and ignores the ways households may have compromised their premove preferences to lower their moving costs. The nature of those tradeoffs can be inferred indirectly from examining how moving costs affect the adjustments of otherwise similar households who face differential costs or they can be examined directly by comparing measures of expressed premove preferences against actual moving outcomes.

## Strengths and Weaknesses of the Search Model

Like all behavioral models, our three-stage search model has both strengths and weaknesses. The model's strengths arise from its advantages over alternative models in identifying and estimating how search and relocation costs affect residential mobility. Those advantages are several. First, the model explicitly recognizes that transaction costs can affect moving behavior in a variety of ways and permits those alternate effects to be identified. Second, this approach enables us to estimate and compare the effects of individual cost items on moving behavior. Finally, this approach provides direct policy leverage by identifying how specific search costs can affect different aspects of the consumption adjustments households make when they move.

The two principal problems with the model are the way it measures the potential benefits of moving and its failure to incorporate measures of housing supply. Unlike the economic models that attempt to derive a single "objective" measure of potential moving benefits, our model proposes to use a series of objective and subjective measures of premove housing circumstances as indirect indicators of the benefits households can expect to receive from moving. Both approaches are subject to problems. The single benefit measure, as noted above, ignores the multidimensional character of housing and the fact that households derive utility from their consumption of specific housing attributes, e.g. space or neighborhood quality, and not from a volume of abstract housing services. Moreover, attempts to expand the objective measure approach to incorporate the multiple dimension of housing rely on untested assumptions about the demographic and economic determinants of demand, the distribution of attribute prices, the identification of a standard population consuming optimal housing, and the appropriate utility function to use in estimating optimal consumption. Without verification of these assumptions, the degree to which such direct benefit measures are truly "objective" is open to question.

The indirect indicator approach, on the other hand, while recognizing the importance of individual housing attributes to a household's potential benefits, fails to provide a direct link between a household's current consumption and the benefits it might realize by moving. This problem may not be particularly troublesome with respect to the household's decision to search or the choice of search procedures, since one could reasonably argue that current levels of stress are more salient to these decisions than some indeterminant perception of potential benefits. However, once a household has actually searched and, consequently has a clearer idea of those benefits, the problem may be more severe. Indeed, Cronin's (1978) findings that dissatisfaction is a better predictor of the decision to search than the direct benefit measure (whereas the reverse is true of the decision to move after a search has been conducted) suggest that may be the case. One could, of course, assume that a household's evaluation of its current housing is implicitly made relative to its perception of the alternatives available in the market. As such, a household's evaluation of its current unit would necessarily reflect, even if at times inaccurately, the potential benefits it could expect from moving ( Adler, 1979 ). Moreover, if, as the model assumes, transaction costs are more likely to influence the type of adjustment made than the decision to move, then the indirect benefit measure should not bias the results.

A second problem with the model is its failure to explicitly incorporate supply variables. In theory, supply conditions can affect search procedures and their effectiveness in at least three ways. First within a particular housing market the appropriate and efficient search procedures may vary by submarket. For example, as suggested above, the appropriate strategy for prospective buyers of expensive single family houses might prove to be a costly and inefficient approach for low-income renters. Second, even within a particular submarket, appropriate search strategies may vary for different households because the suppliers of housing treat such households differently. For example, if minorities or families with children are treated differently by housing suppliers then their search procedures will differ from other households regardless of their other personal characteristics. Finally, varying market characteristics either between markets at a particular time or within a specific market over time may affect search procedures and their outcomes. For example, searchers in a loose market may find that their best strategy is to postpone selecting a unit until they have examined several alternatives while searchers in a very tight market may not have that option.

By omitting supply variables, our model implicitly assumes that all searchers face identical supply conditions and constraints--an assumption that is likely to be unrealistic. However, choosing the appropriate supply variables to include in the model is a difficult task given the diversity of households and housing units in most housing markets. Moreover, previous attempts to capture the influence of supply conditions on moving behavior with vacancy and turnover rate data have not been successful (Goodman, 1978). Nonetheless, an indirect test of potential supply side effects can be incorporated into this model in a number of ways. The presence of submarket effects can be examined by comparing search procedures and their outcomes across different types of units. Similarly, the presence of discrimination by suppliers can be detected by examining the frequency and sources of discrimination reported by searchers. Finally, the influence of overall market conditions can be estimated by comparing results between markets or for the same market at different times.

## SEARCH STRATEGIES AND THEIR OUTCOMES

Although a complete test of this search model has not yet been conducted, we are able to present evidence bearing on three important questions about the search process. First, to what extent do households use different search procedures? Second, how does the nature of the housing market, particularly the behavior of housing suppliers, affect search procedures? Third, do search procedures affect moving outcomes?

The data used to answer these questions were gathered in the baseline surveys of tenants and homeowners in two Midwestern housing markets: Brown County, Wisconsin, whose central city is Green Bay, and St. Joseph County, Indiana, whose central city is South Bend. These surveys were conducted as part of the Housing Assistance Supply Experiment (HASE), a multiyear social experiment conducted by The Rand Corporation under the sponsorship of the Department of Housing and Urban Development. HASE was designed to test the market-wide effects of a full scale housing allowance program on two metropolitan housing markets. A complete description of the Supply Experiment can be found in The Rand Corporation (1979).

## Comparison of Search Strategies

The theoretical and policy significance of housing search to the residential mobility process in predicated on two propositions: first, that the search or shopping procedures households use in locating a new residence differ; second, that differences in search procedures can indeed affect moving outcomes. The rationale for expecting search procedures to vary among households is reasonably straightforward. A household's choice of a residence is a significant consumption decision which is complicated by uncertainty. The significance of the decision is manifest in two ways: first, housing expenses constitute the largest single component of most households' budgets and second, a household's residence shapes a major portion of its living environment. The uncertainty surrounding that choice is due to two factors. First, unlike many consumer products housing is not standardized, instead, it is a multi-dimensional good whose attraction for households depends upon its particular combination of a wide variety of attributes including size, quality, layout, neighborhood and cost. Second, because the transaction costs of changing residences can be substantial, households are not continually in the market for a new residence. Instead, they enter the market relatively infrequently and thus often lack a clear idea of what is available and how much it will cost to change residence. Households adopt search strategies, therefore, to reduce their uncertainty about the costs and benefits of alternative actions. Those strategies necessarily include decision about what information sources to consult and how much effort to expend, decisions that can influence the type of adjustment they make. Because the extent of that uncertainty, the benefits of moving, and the type of unit preferred will all differ among households, we expect search strategies to differ. Moreover, a household's initial plans may change because of information gathered or problems encountered during the search and these experiences will also affect search procedures.

Although our surveys contain no direct measures of search strategies, they can be gauged in several ways, according to the procedures used in the search. Our focus here is on three measures of the effort expended during the search:

(1) the length of that search,
(2) the number of units examined, and
(3) the number and type of information sources consulted.

As the results in Table 2.1 indicate there is indeed considerable variation in observed search behavior in both housing markets. For example, although renters in both markets search, on average, for about two weeks, over one-third search for one week or less while about 20 percent search for over a month. Similarly, while most renters look at only 2 units and about a third choose the first unit they find, another 20 percent examine 7 or more alternatives. Finally, even though most renters rely on newspaper ads and information supplied by friends and relatives, a significant percentage ignore these sources and rely instead on other techniques.

Not surprisingly, owners search longer and examine more units than renters. They also exhibit more variation in their search behavior. The median search length for owners is two months. However, about 20 percent find their units in a week or less while a third spend at least four months in the market. Similarly, more than one-third of all owners examine either substantially fewer or significantly more units than average.

The fact that owners search longer and examine more units than renters is not surprising because the transaction costs of buying and selling a home are substantially higher than the costs of moving into a rental unit. As a result, owners move much less frequently than renters and are less able to correct a bad choice. Moreover, the purchase of a home represents the largest investment most homeowners will ever make. However, the fact that two-thirds of the owners in both sites use agents in contrast to less than one-third of the renters reflects not just differences between the characteristics of prospective owners and renters but also, and, perhaps more importantly, differences between the homeowner and rental markets. Specifically, very few landlords in either housing market use rental agents to find tenants while a substantial portion of all homes for sale are listed with real estate agents.

Without information from the suppliers of housing themselves, it is, of course, impossible to determine the extent to which supply conditions affect households search behavior. However, it is possible to estimate the potential impact of differences in marketing techniques between the renter and ownership markets by comparing the frequency with which renters and owners locate their units with specific sources. This comparison is reported in Table 2.2.

As these data clearly reveal there are marked differences in the rates at which owners and renters locate units with each source. For example, very few renters in either market find units with the assistance of agents, while more owners find their units through agents than in any other way. In addition these data suggest far more clearly the potential impact of different market conditions on searchers behavior than the results presented in Table 2.1. Whereas

Table 2.1: Comparison of Search Effort Among Active Searchers: Renter and Owner Households in Brown County, Wisconsin and St. Joseph County, Indiana

| | Percentage Distribution by Tenure and Site | | | |
|---|---|---|---|---|
| | Renters | | Owners | |
| Search Characteristic | Brown County | St. Joseph County | Brown County | St. Joseph County |
| Length of Search | | | | |
| 1 week or less | 42.6 | 37.9 | 17.9 | 22.3 |
| 1-4 weeks | 39.3 | 34.3 | 20.2 | 18.4 |
| 1-3 months | 13.9 | 18.8 | 25.1 | 29.1 |
| 4+ months | 4.2 | 9.0 | 36.7 | 30.2 |
| Median (days) | 11.3 | 14.8 | 59.2 | 53.3 |
| Alternatives Examined | | | | |
| 1 | 39.5 | 30.5 | 21.1 | 25.5 |
| 2-3 | 18.5 | 26.3 | 12.6 | 7.6 |
| 4-6 | 21.4 | 26.2 | 20.2 | 28.7 |
| 7-11 | 13.0 | 11.5 | 16.4 | 19.7 |
| 12-16 | 4.2 | 2.9 | 11.5 | 8.1 |
| 17+ | 3.4 | 2.6 | 18.2 | 10.4 |
| Median | 2.1 | 2.5 | 6.4 | 5.8 |
| Percent Using Source | | | | |
| Friend or Relative | 70.0 | 82.5 | 79.0 | 68.7 |
| Newspaper Ad | 79.7 | 77.8 | 75.6 | 71.8 |
| Looking at Properties | 31.8 | 46.1 | 64.4 | 74.0 |
| Real Estate or Rental Agents | 23.9 | 27.9 | 68.0 | 66.2 |
| Mean No. of Sources | 2.05 | 2.34 | 2.87 | 2.81 |

*Source:* Tabulated from baseline records of the HASE Survey of Households. Data are weighted to reflect market-wide totals.

Table 2.2 Percentage of Searchers Locating Their Units with Specific Sources

| Source | Renters | | Owners | |
|---|---|---|---|---|
| | Brown | St. Joseph | Brown | St. Joseph |
| Agents | 1.9 | 1.6 | 37.0 | 39.7 |
| Paper | 56.8 | 37.6 | 25.6 | 8.2 |
| Friend | 35.2 | 46.4 | 25.5 | 24.7 |
| Other | 6.1 | 14.4 | 11.9 | 27.4 |
| Total | 100.0 | 100.0 | 100.0 | 100.0 |

the data in the prior table suggested little difference in search behavior within rental and ownership submarkets across sites, the data in Table 2.2 reveal substantial differences between sites--differences that should be expected since Brown County is a virtually all-white community with a tight housing market while St. Joseph County has a substantial and predominately segregated minority population and high vacancy rates.

The renter-owner distinction will not, of course, capture the diversity of housing types and thus the differences in marketing techniques that may be present within a particular market. The average price of homes located through an agent in both counties, for example, far exceeded those found in other ways, $30,000 vs. $23,000 in Brown County and $31,500 vs. $16,200 in St. Joseph County. Similarly the average rent of units found through different sources varies considerably with each market. In Brown County, for example, the few units found through agents had an average monthly rent of $166 vs. $151 for units located through newspaper ads and only $137 for those found through personal referrals. The pattern in St. Joseph was similar viz. $179 (agents) vs. $148 (newspaper ads) vs. $139 (personal referrals). Insofar as these differences reflect distinctly different marketing behavior on the part of housing suppliers they can have a direct effect on the behavior of searchers.

## The Behavior of Housing Suppliers

A clearer picture of why housing suppliers use different marketing approaches and how those approaches affect searchers emerges when we look at data supplied by landlords on their marketing and management policies. Table 2.3 compares the techniques landlords of varying size properties used to recruit tenants in St. Joseph County. These data reveal several important features of landlords' recruiting policies. First, in general landlords do not use many techniques to find tenants. The average number of techniques used is 1.5. Second, landlords in each property size class rely

Table 2.3: Comparison of Techniques Used to Find Tenants Among Landlords in St. Joseph County, 1975

| Recruitment Technique | Number of Units on Property | | | | | All Properties |
|---|---|---|---|---|---|---|
| | 1 | 2-3 | 4-10 | 11-100 | 101+ | |
| Walk-Ins/for rent signs only | 9.4 | 8.4 | 8.3 | 10.9 | - | 9.0 |
| Referrals[a] | 27.5 | 20.9 | 11.6 | 18.2 | 23.1 | 24.4 |
| Newspaper Ads[a] | 31.4 | 39.5 | 45.4 | 14.5 | 15.4 | 34.6 |
| Referrals and Ads[a] | 21.5 | 24.9 | 30.6 | 45.5 | 46.2 | 23.3 |
| Rental Agent[b] | 9.3 | 6.3 | 4.2 | 9.1 | 7.7 | 8.0 |
| T.V. or Radio[b] | 1.1 | - | - | 1.8 | 7.7 | .7 |
| Total | 100.0 | 100.0 | 100.0 | 100.0 | 100.0 | 100.0 |
| Number of Properties | 4,472 | 2,212 | 432 | 54 | 25 | 7,175 |

*Source:* Weighted tabulation from baseline survey of landlords. Comparison limited to landlords who recruited tenants in last year.

*Notes:* a - may also include walk-ins or for rent signs

b - regardless of other technique used

heavily on two sources for finding tenants: personal referrals and newspaper ads. Indeed, 73 percent of all landlords rely exclusively on these two sources--22 percent use only referrals, 30 percent only newspaper ads, and 21 percent both. Third, the number and, to some degree, the types of sources used vary with property size. For example, owners of 1-unit properties use an average of less than 1.5 sources while those whose properties contain more than 10 units use an average of 2.35 sources. Similarly, the mixture of sources used varies with property size. Owners of one-unit properties rely in about equal proportion on either referrals or ads, owners of medium-sized properties (2-10 units) rely more heavily on ads, and owners of large properties tend to use both referrals and ads.

When these patterns are considered in conjunction with the pattern of information source usage among tenants (see Table 2.1), they suggest that tenants rely so heavily on personal contacts and newspaper ads to find units because those are the principal sources landlords use to find tenants. Landlords, in turn, are likely to see at least two advantages to this strategy. First, by limiting the number and types of techniques used, they also limit the costs of finding new tenants. Limiting costs is likely to be particularly important in the St. Joseph County market because the overwhelming majority of landlords there are small scale "mom and pop" operators. (For example, 70 percent of all landlords own only 1 rental property and 45 percent own only 1 rental unit. Moreover, 70 percent spend less than 5 hours per week working on their properties and very few employ a management firm (3 percent) or full or part-time employees (6 percent). Finally, less than 50 percent of all landlords derive more than 5 percent of their income from their rental properties.) Second, landlords are able to tailor their recruiting to their specific needs. Each technique, varies in the degree of exposure it provides for a vacancy and, correspondingly, is the degree to which it allows landlords to screen out tenants. Reliance on personal referrals, for example, will limit exposure to a landlord's friends and relatives and their acquaintances while providing the landlord greater flexibility in screening out undesirable tenants. Newspaper advertisements on the other hand, will provide much greater exposure for a vacancy but make it far more difficult to pre-screen applicants. For the owners of small properties, referrals can provide an inexpensive and effective source of tenants since they have fewer vacancies and correspondingly less need for wide exposure. In contrast, owners of larger properties which in general have more vacancies, will need wider exposure to attract potential tenants even if this requires additional screening checks to sort out undesirable tenants. (Indeed, landlords relying on referrals only use significantly fewer screening checks and are substantially less likely to require leases or deposits than are landlords who use newspapers, rental agents, or advertise on radio or television.)

The importance of personal referrals in filling vacancies in the rental markets also has consequences on differences in renters' search behavior and their outcomes. We have demonstrated elsewhere (McCarthy, 1980) that, contrary to expectations, intensive searches by renters do not yield housing bargains. Instead, searchers who rely exclusively on tips from friends and search for a short time--in essence those who rely on inside information--receive, on average, a 4 percent monthly discount in St. Joseph County and a 6 percent monthly discount in Brown County. Moreover, white searchers relying on inside information in St. Joseph County pay significantly lower deposits than do other searchers. In combination, these results suggest that supply conditions can and do influence household's search behavior.

## Search Procedures and their Outcomes

The significance of differences in search behavior will depend on how search procedures affect the outcomes of the mobility process. We assume that search behavior should have such effects because the success of the search process, like other types of consumer behavior, should depend upon the information available to make housing choices.

There are, of course, several ways to measure outcomes of the mobility process. As we have already noted, in prior work we have focused on the price renters pay for their new residences and whether that price represents a good or a bad deal. We have also examined the move-in costs renters pay when relocating. These cost measures represent only some of the objective measures of outcomes; others would include changes in the total volume of services and changes in specific attributes. All of these objective measures, however, represent near-term outcome measures which reflect specific attributes of the new residences. Movers' assessments of the success of their moves will depend, however, not simply on the attributes of the new residence but also on their perceptions as to whether their new residences meet their housing needs. Since the decision to move to a different residence is indeed a de facto statement of that perception, a near-term measure of satisfaction with a new residence is likely to be biased. Rather, a household's perception of the suitability of its residence, and hence the success of its move, will be better gauged after an interval during which the household has had a chance to become familiar with its new surroundings. Our surveys contain such a measure and we have used it here as a measure of a successful moving outcome.

The specific measure asks whether respondents are more satisfied with their unit now than when they moved in. To estimate the effect of search procedures on this outcome measure responses are compared among owners and renters who actively searched and moved to their units within the past five years (see Table 2.4). The results suggest that search procedures make little difference in the satisfaction of renters but do indeed affect owners' moving outcomes. (These results should be viewed as suggestive since changes in household circumstances as well as search procedures could affect changes in satisfaction.) For example, the differences in the

Table 2.4: Comparison of Initial and Current Unit Satisfaction By Search Procedures Among Recent Movers: St. Joseph County, Indiana

| | More satisfied now than when moved in (%) | |
|---|---|---|
| Search Characteristic | Renters | Owners |
| **Number of Sources Used** | | |
| 1 | 24.8 | 34.6 |
| 2 | 23.8 | 43.0 |
| 3 | 20.8 | 44.2 |
| 4 | 20.8 | 54.2 |
| **Way Unit Found** | | |
| Newspaper ad | 19.0 | 59.1 |
| Real estate or rental agent | 25.3 | 58.6 |
| Looking at properties | 21.4 | 36.9 |
| Friend or relative | 25.6 | 33.7 |
| **Alternatives Examined** | | |
| 1- 3 | 24.7 | 22.3 |
| 4- 5 | 23.1 | 38.4 |
| 6-10 | 13.0 | 61.3 |
| 11-15 | 22.9 | 55.3 |
| 16+ | 52.2 | 43.0 |
| **Length of Search** | | |
| 1 week | 16.6 | 35.5 |
| 2-4 weeks | 20.7 | 38.5 |
| 1-3 months | 26.9 | 36.8 |
| 4-6 months | 21.8 | 53.6 |
| TOTAL | 23.6 | 44.5 |

*Source:* Weighted tabulation from the baseline records of the HASE Survey of households.

percentage of renters reporting that they are more satisfied with their units now than when they moved are small and inconsistent across each of the four search dimensions. In contrast, there are marked differences among owners. Specifically, owners appear more satisfied, the more sources they used, the longer they search and if they used either a newspaper ad or a real estate agent to locate their unit. Interestingly, satisfaction increases with units examined up to a point and then begins to decline suggesting that searchers who are unable to find a suitable residence after examining a half-dozen to dozen residences may be settling for less than they wanted.

## SUMMARY

Our brief review of the evidence suggests the following tentative conclusion. First, housing search behavior does vary among households, although that variation is greater among owners than renters. Second, supply condition, particularly the ways in which housing suppliers recruit prospective buyers or tenants, are likely to affect search procedures and may indeed help to explain why renters' search procedures are less varied than owners'. Third, search procedures appear to affect search outcomes although not necessarily in the same way for owners and renters. For owners, more thorough searches, that is those using many sources, examining a range of alternatives, and lasting several months, appear to increase the chances for a successful outcome. Among renters, search outcomes also vary with search procedures but not in the same way. For renters, whom you know appears more important than how hard you search in determining the outcome of the process--a finding reflected in the importance of inside information in renter's locating bargains and paying lower move-in costs.

## REFERENCES

Abt Associates (1978) "Draft Report on the Search Behavior of Black Households in Pittsburgh in the Housing Allowance Demand Experiment", Unpublished report. Cambridge: Massachusetts.

Adler, N. E. (1979) "Decision models in population research," Journal of Population, 2, 187-202.

Barresi, C. (1968) "The role of the real estate agent in residential location", Sociological Focus, 1, 59-71.

Barrett, F. (1973) Residential Search Behavior. Toronto: York University, Research Monograph.

Brown, L. A. and E. G. Moore (1970) "The intra-urban migration process: a perspective", Geografiska Annaler, 52, 1-13.

Brummell, A. C. (1979) "A model of intraurban mobility", Economic Geography, 55, 338-352.

Butler, E. W., et. al. (1969) Moving Behavior and Residential Choice, Washington, D.C.: National Cooperative Highway Research Program, Highway Research Board, Report No. 81.

Clark, W. A. V. and M. Cadwallader (1973) "Locational stress and residential mobility", Environment and Behavior, 5, 29-41.

Cronin, F. J. (1978) "Intra-urban household mobility: the search process", Washington, D. C.: The Urban Institute, Unpublished paper.

Da Vanzo, J. (1976) Why Families Move: A Model of the Geographic Mobility of Married Couples. Santa Monica, California: The Rand Corporation, R-1972, DOL.

Goodman, J. L., Jr. (1976) "Housing consumption disequilibrium and local residential mobility", Environment and Planning A, 8, 855-874.

----- (1978) "Housing market determinants of movers' city/suburban location choice", Washington, D. C.: The Urban Institute, Working Paper 1014-1.

Greenwood, M. J. (1975) "Research on internal migration in the United States: a survey", Journal of Economic Literature, 13, 397-433.

Hanushek, E. A. and J. M. Quigley (1978) "An explicit model of intra-urban mobility", Land Economics, 54, 411-429.

Hempel, D. G. (1969a) The Role of the Real Estate Agent in the Home Buying Process. Storrs, Connecticut: University of Connecticut, Real Estate Report No. 7, Center for Real Estate and Urban Economic Studies, School of Business Administration.

----- (1969b) Search Behavior and Information Utilization in the Home Buying Process. Storrs, Connecticut: University of Connecticut, Center for Real Estate and Urban Economic Studies, School of Business Administration.

Huff, J. O. and W. A. V. Clark (1978) "Cumulative stress and cummulative inertia: a behavioral analysis of the decision to move", Environment and Planning A, 10, 1101-1119.

McCarthy, K. F. (1979) Housing Search and Mobility. Santa Monica, California: The Rand Corporation, R-2451-HUD

----- (1980) Housing Search and Consumption Adjustment. Santa Monica, California: The Rand Corporation, P-6473.

Meyer, R. (1980) "A descriptive model of constrained residential search", Geographical Analysis, 12, 21-32.

Morgan, J. N. (ed.) (1972) Five Thousand American Families: Patterns of Economic Progress, Vol. II. Ann Arbor, Michigan: Survey Research Center, Institute for Social Science Research.

Newman, S. J. and G. J. Duncan (1979) "Residential problems, dissatisfaction and mobility", American Planning Association Journal, 45, 154-166.

Quigley, J. M. and D. H. Weinberg (1977) "Intra-urban residential mobility: a review and synthesis", International Regional Science Review, 1, 41-66.

Rossi, P. (1955) Why Families Move. Glencoe, Illinois: Free Press.

Silk, J. (1971) Search Behavior: General Characterization and Review of Literature in the Behavioral Sciences, University of Reading, England: Geographical Papers No. 7, Department of Geography.

Smith, T. R., W. A. V. Clark, J. O. Huff and P. Shapiro (1979) "A decision-making and search model for intraurban migration", Geographical Analysis, 11, 1-22.

Speare, A. Jr. (1974) "Residential satisfaction as an intervening variable in residential mobility", Demography, 11, 173-188.

----- S. Goldstein and W. Frey (1975) Residential Mobility, Migration and Metropolitan Change. Boston: Ballinger.

The Rand Corporation (1977) Third Annual Report of the Housing Assistance Supply Experiment, Santa Monica, California: R2151 - HUD.

U. S. Bureau of the Census (1966) "Reasons for Moving: March 1962, to March 1963", Washington, D. C.: U. S. Government Printing Office. Current Population Reports, Series P-20, No. 154.

Weinberg, D. H. (1979) "The determinants of intra-urban household mobility", Regional Science and Urban Economics, 9, 219-246.

----- and R. Atkinson (1979) "Place attachment and the decision to search for housing", Growth and Change, 10, 22-29.

----- et. al. (1981) "Intraurban residential mobility the role of transaction costs, market imperfections, and household disequilibrium," Journal of Urban Economics, 9, 332-348.

Wolpert, J. (1965) "Behavioral Aspects of the Decision to Migrate", The Regional Science Association Papers, 15, 159-169.

# 3

# SEARCH ADJUSTMENT IN LOCAL HOUSING MARKETS

**Gavin Wood and Duncan Maclennan**

Microeconomic analysis of housing issues is generally undertaken within the context of neoclassical or Marshallian approaches to consumer choice and market behaviour. In the static equilibrium version of these models, perfectly informed consumers react immediately, by adjusting price bids or consumption patterns, to market information which is, by assuming all units of a commodity sold to be homogeneous, all encapsulated in price signals generated by the bidding process. As a result, in this paradigm, applied economists are concerned with outcomes, assumed to reflect equilibrium states of the system, which maybe summarised in conventional price and income elasticity measures. This paradigm, which still dominates applied economic research, is thus focussed on market outcomes and individual choice or adjustment processes are of no interest. This lack of concern arises because, by assumption, only "well-behaved" markets are under consideration and because markets are populated by familiar, well-informed, frequent and regular traders who are immediately and directly informed by price signals.

Faced with this conventional microeconomic model of consumer behaviour the economist is faced with two choices. First, it can be assumed that the model reasonably characterises individual and market behaviour and ways are sought by which further implications or scope of the model maybe pursued. (Arguably, by focussing interest on the neoclassical general equilibrium model, theoretical economists have made an inordinate effort to develop general models from this specific notion of commodities, individuals and markets.) The second alternative, is to ask whether markets and individuals do indeed function in a fashion even remotely similar to the theoretical model. That is, based on what is known about markets, are the models reasonable.

It is our contention, and indeed the justification of our subsequent empirical sections, that static neoclassical models of consumer demand do not represent a plausible framework in which to analyse housing choice decisions. This is a strong statement and it implies that a less restrictive approach is required for applied housing economics. There are characteristics of housing and the housing


DOI: 10.4324/9781003182085-4

market which make a search adjustment/information acquisition perspective an appropriate starting point.

## THE HOUSING MARKET - RELEVANT FEATURES

We define the housing market search process as comprising those activities which aid the potential mover's acquisition of knowledge pertinent to the successful attainment of aspirations. There are a number of housing market characteristics which imply that analysis of this search process is crucial to an understanding of the housing choice decision. Such salient features are:

(1) Individuals transact in the housing market infrequently. Whilst non-moving households may acquire a broad, continuous stream of housing market information, from the mass media or friends transacting in the market, the specific information required for purchase will only accrue with purposive search. The average household transacts in the market only once every seven years in Britain. More rapid mobility rates are only commonplace in the furnished rental sector and for the early life cycle stages of the household. As a result the quality of information stored in the individual's memory is likely to decay over time. That is, there is a "forgetting" curve.

(2) The problem of consumer memory loss is compounded by changes in the housing market in the non-moving period. With long periods of non-transaction, the dynamics of the housing market may alter the quality, neighbourhood characteristics, relative accessibility and relative prices of large areas of the housing stock. These changes may render obsolete that information which the household gleaned from past market participation. Such a tendency will be reinforced by instability in the evolution of the market, as encouraged by periods of disequilibrium. However, even if the broad characteristics alluded to above have been monitored by the potential movers, detailed information on property characteristics, prices and vacancy rates can be elusive. For instance, in the privately rented sector the probability of finding a vacancy, of a given type in a particular area, may change on a daily basis (Maclennan, 1979a) and in the owner occupied sector recent British experience (in 1973 and 1978) has shown that house prices may rise by 10-15 percent in a three month period and over such a period relative prices within a single housing market may become distorted across sub-markets. It would appear appropriate to conclude that the rapid redundancy of price and availability information in the housing market combined with the

infrequency of purchase or movement will generally result in potential movers having imperfect information.

(3) Points 1) and 2) provide a rationale for why market information is likely to be poor at the onset of search activities. Further, it can be noted, that the burden of updating that information via search is exacerbated by the spatially dispersed nature of housing vacancies. Particularly in larger urban housing markets, the spatial separation of purchasing opportunities adds to the time, travel and psychic costs of housing search.

(4) Search activities are particularly desirable if the housing market is in disequilibrium. In the circumstances of the housing market, disequilibrium evolves in the form of a mismatch between the pattern of demand and supply. In the absence of instantaneous adjustment, the households' chances of attaining housing aspirations depend not just upon ability and willingness to pay, but also upon the ability to successfully negotiate in the market. Clearly, the stock of market knowledge is pertinent here. There is good reason to believe that disequilibrium may be pervasive. The fixity of the second-hand housing stock and its relatively slow rate of turnover (partly related to recontracting costs) is likely to mean that submarket demands may change more rapidly than supply. Further, lagged new supply completions and the possibility of disequilibrium in the market for housing finance are all likely to contribute to housing market instability.

(5) Though we have argued that market participants will find it desirable to improve their stock of knowledge, the search costs referred to above could conceivably be of a size which militate against engaging in any purposive search activities. There are two features of the housing market which lead to the rejection of this suggestion. Firstly, housing represents a major consumption and investment good purchase for most households; thus in general, the benefits to search will be greater than with regard to most other household good purchases. We can then expect the consideration and evaluation undertaken by households to be more intense with regard to housing. Secondly, this tendency is reinforced by the financial barriers to the redeeming of a "false trade" by instant re-contracting. These financial barriers are the search and transaction costs which are likely to make re-contracting and movement expensive. Including legal fees and removal costs these costs may range between 5 and 10% of the total price of a house, particularly where movement entails both selling and purchase costs.

(6) Having established search activities as an important component of the process of purchase, there now

remains the issue of what features characterise these activities. Search could be of a passive character with, for instance the household making the purchase decision solely upon the acquisition of messages which are relayed impersonally, but which describe important features of vacancies. Alternatively, search could be of an active character in which the household personally evaluates the attributes of vacancies by direct observation. Housing is in fact a commodity which fails to fit neatly these categories. Clearly, the aesthetic qualities of housing/neighbourhoods are aspects which can only be judged to the satisfaction of the household by personal evaluation, and in this regard we can expect households to exhibit resistance to a reliance upon messages relayed impersonally. However there are of course technical and complex housing attributes, such as structural stability for instance, for which households' evaluations must rely upon the assessments acquired from "experts" such as surveyors, building societies or estate agents. So both personal consideration and interaction with housing market "professionals" can be expected to characterise search activities.

(7) One important aspect of the house purchase process which has not as yet received attention with regard to search activities, is that the potential entrant is required not only to evaluate dispersed housing offers and to successfully pursue (in the owned sector) a search for loan finance, but also to place bids. The actual form of the housing market transaction may vary over space and time. For instance buyers may personally respond to a fixed price offer of sale (particularly in a buyers' market) or they may enter a direct bargaining process with the seller. If such bargains are agreed they may be legally binding or they may be informal. It is the latter characteristic of the English pricing system that gives rise to gazumping when buyers or sellers renege on non-binding contracts. In the Scottish system sales generally take place by presenting sealed bids to the sellers agent, with the latter usually accepting the highest bid. Open auctions of housing are now uncommon. Thus, the generally prevailing systems of selling housing, and its associated search effort, conform neither to a well behaved Marshallian market nor a Walrasian auction.

However, even taking into account all these pervasive characteristics of the housing market, which make search inevitable, a general deterministic theoretical model cannot be convincingly developed. This is because the nature of housing search, the role of housing market institutions and the characteristics of information and bidding systems will vary across different contexts. For instance, in the renter sample examined in this paper, rent controls had dictated that households were primarily concerned with establishing the existence of housing vacancies across a narrow range of housing price and quality variation. In the owner sample, households had to search jointly for housing and house purchase finance. And, of course, owner searchers were concerned to make relatively long term asset purchases thus implying that a detailed evaluation prior to purchase would be essential.

## SEARCH ADJUSTMENT IN HOUSING MARKETS

The particular interest of this chapter which is a companion paper to one later in the book is the housing market entry process. Viewed sequentially, the entry process can be partitioned into factors triggering the decision to search, the search process itself and the outcome in terms of a relocation decision (McCarthy, 1980). It is plausible to argue that the search process itself can be an important influence upon housing choice (McCarthy 1980) and it is this specific aspect of the entry process with which this chapter is most concerned. In particular, attention is devoted to the problem of modelling linkages between search activities and outcomes. Two features characterise our approach here; firstly, use is made of conventional techniques of the economic analysis of choice, but constraints are adapted to take account of the conditions of imperfect information and disequilibrium which entrants operate under. Secondly, assumptions are tailored to conform with empirically based generalisations on household behaviour. This latter point reflects the importance of spatial and temporal variations in housing market conditions and institutions.

The empirical work presented in this chapter is based upon an investigation of the City of Glasgow housing market. The character of the entry process differs considerably across tenures, and hence our analysis of private rental tenants and house purchasers is presented separately. In the next section we examine the private rental sector and the following section is devoted to the owner occupied sector.

## SEARCH AND OUTCOMES IN THE PRIVATE RENTAL SECTOR

In the period 1974-1976 the University of Glasgow conducted an economic analysis of student and furnished rented housing in the city of Glasgow. It became apparent that the sector, as in many other British cities, was characterized by the transient nature of tenants in furnished lets and by indications of excess demand for properties in the sector. (See also, Maclennan, 1977 and 1978). Within this tenure sector students are an important and distinctive group. Some 40 percent of tenants in the sector in Glasgow were students and their housing demands were characterised by their intermittent and short run nature. At the end of each academic year a large, if declining, proportion of students leave their rental addresses. With the start of the next year there is then a peaked inflow of returning students and new students which creates substantial pressures in the local housing market. In this study it was possible to ascertain, from three sources, the time distribution of the flow of student searchers into the local housing market. This flow for 1974 was not significantly different (at the 0.01 level) from the equivalent distributions for 1971, 1972 and 1973. Proxies for weekly vacancy flows, derived from newspaper advertisements indicate that the vacancy flow has a relatively smooth annual pattern with higher levels in the early summer periods possibly reflecting annual student quitting (Maclennan, 1978). It can therefore be argued that excess demand for furnished sector vacancies also has a marked peak and a proxy for the state of market tightness is shown in Figure 3.1. Excess demand

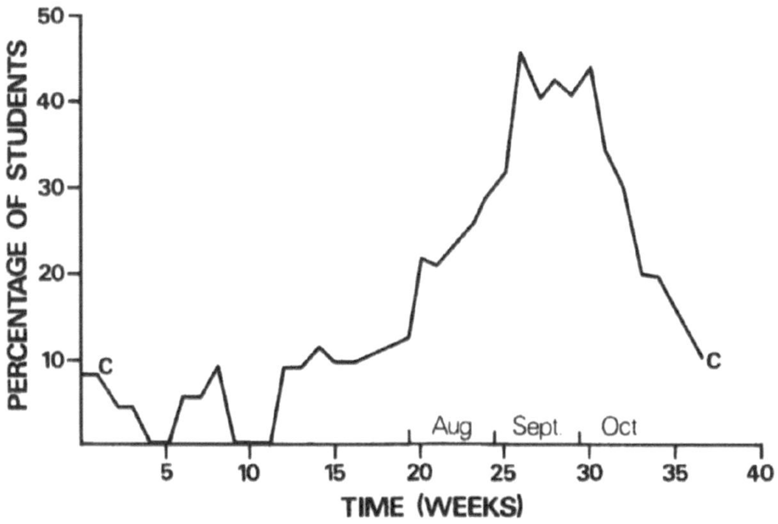


Figure 3.1. Excess demand proxy: percentage of students searching in 1974. Source: Maclennan (1979b).

for furnished lets at time $t_n$ can be indicated as

$$E_{t_n} = A_{t_n} - V_{t_n} + \left(\sum_{t=0}^{n-1} A_t - V_t\right) \qquad (1)$$

where E is excess demand, $A_{tn}$ arrivals at time $t_n$ and $V_{tn}$ vacancies at period $t_n$. The line CC, in Figure 3.1 is an estimate for students in Glasgow of the sum of current arrivals to search plus past arrivals still unhoused, that is:

$$CC = A_{t_n} + \left( \sum_{t=0}^{n-1} A_t - V_t \right) \tag{2}$$

and since $V_t$ was relatively invariant then CC is a reasonable indicator of excess demand in the local housing market. However student housing preferences are not only bunched in time but also residential aspirations are concentrated spatially. A previous paper has indicated (Maclennan, 1977) that students generally aspire to reside near the local university as their work and play activities are concentrated around the campus. The focus of this section is to explain how the peak of arrivals and pattern of locational aspiration are transformed respectively into the time distribution of successful searches and the actual pattern of residential locations. The two problems are closely related since there are systematic spatial variations in both the level of excess demand for lets and also expectations about the pattern of excess demand through time and over space.

As a theoretical basis for empirical investigation the access-space trade off model would appear to be unsuitable. There are two main reasons for arriving at such a conclusion. Firstly, empirical analysis of the patterns of rents and housing quality around the University over a four-year period indicated that there was no obvious rent gradient. Secondly, rent controls were present thoughout the sample period. Hence spatial rationing by a rent bidding process is precluded. A non-price allocation mechanism therefore operates.

Rather than operating randomly, the allocation mechanism can be viewed as being based upon the search procedures of individuals operating under imperfect information. This belief is reinforced by the response of 1200 students entering the private rental sector in Glasgow and surveyed during the period 1974-75; they were asked to identify and rank the major difficulties of finding and living in furnished rental housing. The majority of students consistently indicated that the direct, psychic and time costs of searching for housing was their major "housing problem".

Data from this survey is used here to derive empirically based simplifying assumptions for the purpose of constructing a model of housing search. These have been reported on elsewhere (Maclennan, 1979a; 1979b) and here we merely summarise the generalisations which could be established:

(1) There was substantial variation in the information levels possessed by students at the onset of search. In particular, individuals who had previous experience in the housing market were more capable in identifying "efficient" search areas (efficiency as indicated by a vacancy density measure) than were newcomers.

(2) Variance in information levels was complemented by differences in expectations regarding the spatial pattern of excess demand. In particular, newcomers exhibited difficulty in articulating expectations, while those with

firm perceptions revealed a broadly declining expected excess demand with distance from the university. Experienced students reveal more sophisticated anticipations in that excess demand was expected to exhibit a pronounced peak in areas adjacent to the university while matched by a pronounced trough toward the edge of the city.

(3) For two sub-groups of the sample, local and non-local students, there was considerable difference in the costs of search. For most students the major element of search costs is the opportunity cost of abandoning other activities, particularly summer employment, to arrive to search for housing. Next most important are the financial and educational costs of living in temporary accommodation, particularly hotels, until term accommodation is obtained. Clearly local students bear neither of these costs. They can search (if less efficiently) whilst maintaining summer activities and they can live at home until an appropriate vacancy is established. The costs of such local searching are, in financial terms, phone calls and travel costs to the city which may be spread over a searching group of three or four fellow students who form a common searching group. The indirect cost of consuming "home" rather than "flat" consumption services may be high but is equally incurred by non local groups. The latter individuals have the problem of paying for a temporary searching base. Students returning to University may be able to live, informally, temporarily and cheaply with past acquaintances, thus lowering their financial search costs. Completely new, non local students do have the problem of living in high cost or low quality short'term accommodation which is frequently unsuitable for home study purposes. Students in this category not only have a problem of financial cost but they tend to see the start of term as being a constraint date by which they must have term time accommodation.

(4) A further empirical observation greatly simplified the construction of a search model. The available evidence indicated that students accepted the first vacancy offered. This simple observed stopping rule indicates in the market being examined, wherein excess demand was prevalent and there were no marked variations in property rents/quality characteristics, that search is intended to establish the existence of vacancies rather than provide a comparative valuation of the price/quality characteristics of a known set of vacancies. Further, letting was generally conducted on a "first-come-first-served" basis and offers could not be stored without considerable cost.

## A Model of Search Distance and Outcomes

The search adjustment process can be simply illustrated by making use of the following assumptions. Firstly, students select an initial search location in accord with their aspirations. If unsuccessful, the information gleaned from the failure is used to add to and clarify expectations regarding the spatial pattern of excess demand, expected duration of search (given search location) and search costs, which are assumed to be dependent upon duration of search. Secondly, students adopt a "first vacancy accept" stopping rule, given a maximum intended search cost constraint.[1] The problem is then to revise distance and duration of search so as to ensure an expected successful outcome compatible with the search cost constraint. This can be expressed and summarised in the following form:

$$\hat{V} = \hat{V}\,(\bar{N},\ A) \qquad d\hat{V}/dA > 0 \qquad (3)$$

$$D = D\,(\hat{V}) \qquad dD/d\hat{V} < 0 \qquad (4)$$

$$Sc = Sc\ (D) \qquad dSc/dD > 0 \qquad (5)$$

$$Sc = \widetilde{Sc} \qquad \widetilde{Sc} > 0 \qquad (6)$$

where $\hat{V}$ is probability of vacancy acquisition and $\bar{N}$ is number of searches per time period, which, for purposes of simplification, is assumed constant. A is distance from work place, and positively influences $\hat{V}$. D = expected duration of search measured in terms of time periods and is negatively related to $\hat{V}$. Sc = search costs which are posited as a function of search duration. $\widetilde{Sc}$ = maximum intended search costs. To analyse the implications of this model we initially adopt the following linear specifications of (3) - (6).

$$\hat{V} = \delta\bar{N} + \xi A \qquad 0<\delta<1, \qquad 0<\xi<1 \qquad (7)$$

$$Dz = \bar{X} - \phi V \qquad \phi > 0 \qquad (8)$$

$$Sc = \alpha D \qquad \alpha > 0 \qquad (9)$$

$$Sc = \widetilde{Sc} \qquad (10)$$

By substituting (7), (8) and (9) into (10) the values $\widetilde{V}$, $\widetilde{A}$ and $\widetilde{D}$ which satisfy the search cost constraint, can be derived. Thus

$$\widetilde{Sc} = \alpha\bar{X} - \phi\alpha(\delta\bar{N} + \xi A) \qquad (11)$$

The value of A is then:

$$\widetilde{A} = \frac{\bar{X}}{\phi\xi} - \frac{\delta\bar{N}}{\xi} - \frac{\widetilde{Sc}}{\alpha\phi\xi} \qquad (12)$$

Substituting (12) into (7) yields $\tilde{V}$

$$\tilde{V} = \frac{\bar{X}}{\phi} - \frac{\tilde{S}c}{\alpha\phi} \tag{13}$$

Substituting (13) into (8) gives $\tilde{D}$

$$\tilde{D} = \frac{\tilde{S}c}{\alpha} \tag{14}$$

It has already been noted that searchers possess different past housing histories and hence reveal differing expectations and search costs. In order to derive empirically testable hypotheses from the model we therefore make use of the generalisations introduced earlier. Let us first investigate differences in search costs. Entrants who are newcomers and lack past experience in the housing sector, can be expected to experience higher search costs. In terms of the postulated model, we are interested in the response of $\tilde{A}$, $\tilde{V}$ and $\tilde{D}$ to different values of the parameter $\alpha$. The signs of the partial derivatives $\partial\tilde{A}/\partial\alpha$, $\partial\tilde{V}/\partial\alpha$, $\partial\tilde{D}/\partial\alpha$, indicate the expected direction of change:

$$\frac{\partial\tilde{A}}{\partial\alpha} = \frac{\tilde{S}c\,(\phi\xi)}{(\alpha\phi\xi)^2} > 0, \tag{15}$$

$$\frac{\partial\tilde{V}}{\partial\alpha} = \frac{\tilde{S}c}{(\alpha\phi)^2} > 0, \tag{16}$$

$$\frac{\partial D}{\partial\alpha} = \frac{-\tilde{S}c}{\alpha^2} < 0, \tag{17}$$

Hence, in order to accommodate a higher $\alpha$ newcomers will expect to make successful choices at less accessible locations, with a higher probability of acquisition and lower duration of search.

By allowing for differing expectations across subgroups further implications can be deduced. Pertinent here is our earlier observation that experienced searchers expect excess demand to exhibit a pronounced peak in areas adjacent to workplace and a pronounced trough towards the edge of the city. This feature can be built into the model by altering either the accessibility-vacancy probability specification or the vacancy probability - duration of search specification, i.e. we presume that differing expectations regarding the spatial pattern of excess demand will be matched by commensurate variation in D for any given $\hat{V}$ and similarly in $\hat{V}$ for any given A. To illustrate this particular adaptation of the model we choose to respecify (8) as:

$$D_E = \bar{Y} - \phi\tilde{V}_E^2 \tag{18}$$

where the subscript E denotes the subgroup of experienced searchers. Denoting the remaining equations as

$$\hat{V}_E = \delta\bar{N} + \xi A_E \tag{19}$$

$$Sc_E = \alpha D_E \tag{20}$$

$$Sc_E = \tilde{S}c \tag{21}$$

then solution values for $\tilde{A}_E$, $\tilde{V}_E$ and $\tilde{D}_E$ can be derived from the following expression for Sc

$$\tilde{S}c = \alpha Y - \alpha\phi[\delta\bar{N} + \xi A]^2 \tag{22}$$

$$\tilde{A}_E = \frac{1}{\xi}\left[\sqrt{\frac{\alpha\bar{Y} - \tilde{S}c}{\alpha\phi}} - \delta\bar{N}\right] \tag{23}$$

$$\tilde{V}_E = \sqrt{\frac{\alpha\bar{Y} - \tilde{S}c}{\alpha\phi}} \tag{24}$$

$$\tilde{D}_E = \frac{\tilde{S}c}{\alpha} \tag{25}$$

Notice that with an identical specification for (20) and (21), duration of search will be the same for both groups. $\tilde{A}_E$ and $\tilde{V}_E$ can however differ. Given identical durations then $\tilde{A}_E < \tilde{A}$ if $\tilde{V}_E < \hat{V}$, i.e. if

$$\sqrt{\frac{\alpha\bar{Y} - \tilde{S}c}{\alpha\phi}} < \frac{\bar{X}}{\phi} - \frac{\tilde{S}c}{\alpha\phi} \tag{26}$$

Since the value of D is common to both groups, the inequality condition can be expressed as

$$\sqrt{(Y-\tilde{D})/\phi} < (X - \tilde{D})/\phi \tag{27}$$

which implies that the values of Y and X are crucial, and further, the inequality could reverse in sign at different values of D. The four quadrant figure 3.2 illustrates such a case. The line $D_E D_E$ is that which most closely approximates the expressed spatial expectations of experience searchers with $\bar{Y} < \bar{X}$. Figure 3.2 presents the case where $\tilde{S}c$ (quadrant a) is at a level where solution values for V and A are such that experienced searchers paradoxically choose higher values for both variables; in particular, experienced searchers have an expected duration of search of $\tilde{D}$ (quadrant c) and an accessibility given by the $A_3$ vacancy probability schedule (quadrant b), with $A_3$ greater than $A_2$ the choice of newcomers. Clearly, if the search cost constraint allowed expected duration $>\hat{D}$, then the inequality in solution values would be reversed for the two subgroups. Figure 3.2 also depicts the influence of changes in the value of the parameter $\alpha$ across subgroups. $Sc_E$ (quadrant d) indicates the lower

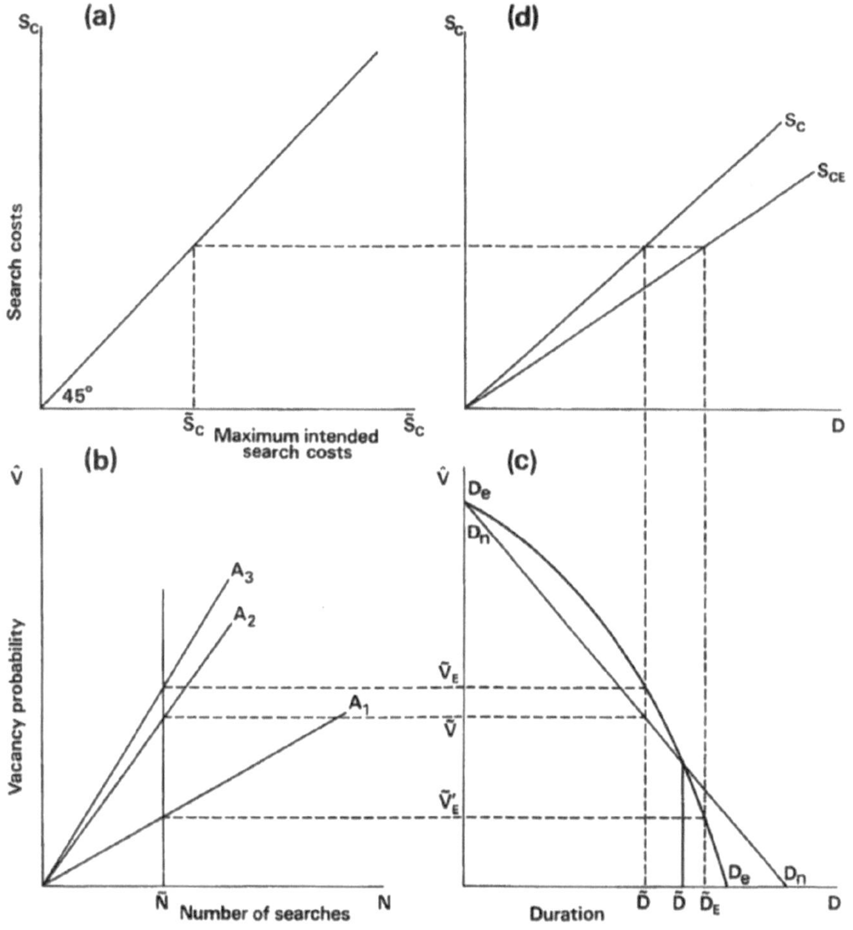


Figure 3.2. The interrelationship of search costs, duration of search, and number of searches.

search costs incurred by experienced searchers. As deduced earlier, the unambiguous result is that of increasing duration, from $\tilde{D}$ to $\tilde{D}_E$, and improvement in accessibility to work place from $A_3$ to $A_1$. The following hypotheses can therefore be subjected to empirical investigation. For experienced searchers with lower search costs then, ceteris paribus:

Hypothesis 1: The successful search distance will be closer to work place.

Hypothesis 2: The duration of search will be longer.

Use of the assumption that all potential entrants select an initial search location in accord with aspirations allows us to deduce implications concerning the search adjustment process. Figure 3.3 illustrates this process. Once again the outcome for experienced/

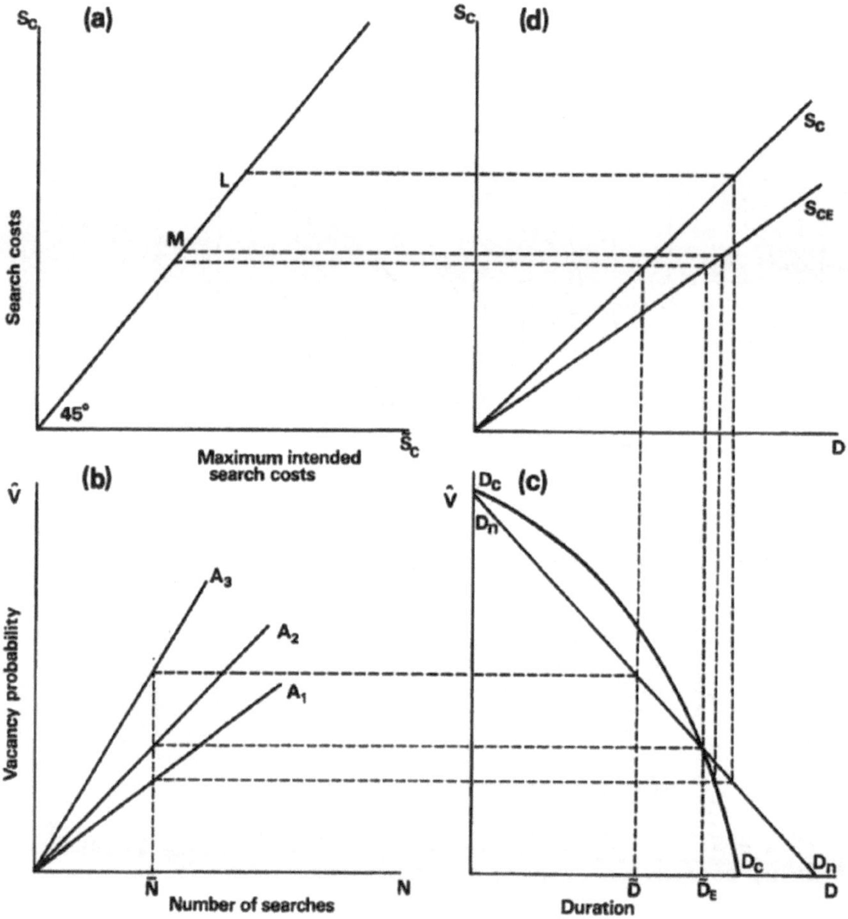


Figure 3.3. The search adjustment process.

newcomer subgroups is ambiguous as a consequence of the countervailing influence of differences in search costs and vacancy probability duration relationships. In the particular instance illustrated by figure 3.3, the former outweigh the latter and we are left with the plausible outcome that newcomers are forced to adjust over distances greater than experienced searchers.

This result can be traced out in figure 3.3 by presuming that both newcomers and experienced subgroups make an initial search at the identical distance $A_1$ (quadrant b). Given an identical search cost constraint $\widetilde{Sc}$ (quadrant a) then, on the basis of expectations clarified by this initial search (as summarised by the relationships in quadrants b-d), it is evident that expected search costs at M and L are greater than $\widetilde{Sc}$ for both experienced and newcomer searchers respectively.[2] Searchers must therefore adjust search duration and location. The adjustment which satisfies the expected search cost constraint involves a choice by newcomers of distance $A_3$ and duration $\widetilde{D}$, and a choice by experienced searchers of distance $A_2 < A_3$ and duration $\widetilde{D}_E > \widetilde{D}$. The following hypothesis relating to the search adjustment process can therefore be stated. For experienced searchers with lower search costs then ceteris paribus:

Hypothesis 3: The adjustment to search distance will be less rapid.

Tests of Hypotheses

The hypotheses deduced above related to subgroups differing with respect to their past housing history, i.e. experience in the housing market. The categories from our student sample compatible with this distinction are,

1) New-entrant and/or non-local students
2) Continuing and/or local students

The previous experience and hence market information of the latter group can be expected to be substantially greater than the former, and their search costs relatively smaller. Table 3.1 presents results on mean successful distance and mean duration of search for these subgroups. They are compatible with hypotheses 1 and 2 above in that they suggest that the, on average, greater search costs incurred by subgroup 1 outweigh any countervailing influence of differing expectations concerning the relationship between vacancy probability and expected search duration.

Table 3.1: Average Search Measures

| | New entrant and/or non-local students | Continuing and/or local students |
|---|---|---|
| Mean successful distance (km) | 3.67 | 2.84 |
| Mean search duration (days) | 7.40 | 18.50 |

Search adjustment in housing markets

Multiple regression analysis was employed in order to investigate hypothesis 3 concerning the adjustment process of entrants. The variables used in the test were:

| | |
|---|---|
| Ai | The percentage change in search distance between search i and search i-1. |
| Di-1 | Search distance, measured in terms of Km from university. |
| G | Size of search group. |
| Y | Income level of searcher. |
| F | An excess demand proxy for the week of search. The estimate is provided by calculation of the expression $E_{tn}$ in equation (1). |
| L | Local/non-local dummy value. L takes the value unity if a local student and zero otherwise. |
| U | Error term. |

A number of comments are pertinent here on the method by which hypothesis 3 is tested, and the independent variables considered relevant. The regression model, equation (28), was estimated separately for new-entrant and continuing groups, while origin of student was distinguished by use of the dummy variable L. Place of origin is likely to exert the strongest influence upon differences in search costs, while the new/continuing entrant distinction is more closely related to differing market expectations. Given relatively homogeneous market expectations within the new/continuing subgroups, then the dummy variable L can be expected to capture the influence of differences in search costs. One qualification to this statement is recognised by the inclusion of G. The burden of search costs can be mitigated by spreading costs across a group of individuals acting in concert.

The variable F is included in recognition of the temporal variation in excess demand, which in turn can be expected to alter the spatial pattern of searcher expectations. Our analytical model assumed that expectations are "frozen" following the initial search. As we noted earlier there is in fact substantial temporal variation in expected excess demand. There remains the variables Di-1 and Y. The former is included to standardise for the statistical influence of the distance of previous search, exerted as a consequence of it being, as the dependent variable, specified in terms of percentage change. Furthermore, as the students aspire to locate near their workplace, then any variance exhibited by these initial aspirations will influence size of subsequent adjustment, given expectations concerning the spatial pattern of excess demand. Lastly, Y is included on the grounds that if students exhibit a similar propensity to devote funds from income to search, then the search cost constraint will vary correspondingly. The following linear specification was estimated by O.L.S. and the results given in Table 3.2.

$$A_i = b_o + b_1 D_{i-1} + b_2 G + b_3 Y + b_4 E + b_5 L + U \qquad (28)$$

Examination of the results for the new-entrant group indicates that the overall level of statistical explanation was generally higher than for continuing students (Table 3.2). The explanatory power of the equation and the overall significance vary interestingly as search frequency is increased.

Table 3.2: Search Adjustment -- Regression Results

| Adjustment | $D_i - 1$ Distance i - 1 | G Group Size | Y Income | E Excess Demand | L Local | $R^2$ |
|---|---|---|---|---|---|---|
| Newcomers | | | | | | |
| 1 | -0.07* | -0.03 | 0.002 | 0.03* | 0.24* | 0.29 |
| 2 | -0.12* | -0.09 | 0.001 | 0.02* | 0.16 | 0.26 |
| 3 | -0.84* | -0.14* | -0.003 | 0.008 | 0.53 | 0.41 |
| 4 | -0.77* | -0.09* | ----- | 0.003 | 1.20* | 0.54 |
| 5 | -0.31* | -0.16* | 0.002 | 0.009 | 4.78* | 0.51 |
| 6 | -0.22* | -0.15* | 0.002 | 0.002 | 1.38 | 0.49 |
| Experienced students | | | | | | |
| 1 | +0.03 | -0.07* | 0.001 | 0.14* | 0.18 | 0.18 |
| 2 | -0.09 | -0.11* | -0.004 | 0.09* | 0.21 | 0.23 |
| 3 | -0.12 | -0.10* | 0.003 | 0.11* | 0.32* | 0.34 |
| 4 | +0.08 | -0.12* | 0.002 | 0.12* | 0.71* | 0.37 |
| 5 | -0.02 | -0.17* | 0.081 | 0.14* | 0.84* | 0.40 |
| 6 | -0.17* | -0.14* | 0.092 | 0.08* | 1.03* | 0.42 |

*Significant at the 0.05 level.

For the first two searches by newcomers the coefficient on Di-1 is statistically significant and close to zero. However, since students exhibited little variation in initial aspirations (in terms of accessibility) this is not altogether surprising (see Maclennan, 1979:259). Hence significant non-uniformity in previous search location emerges after the initial unsuccessful search has been experienced; this is reflected in the increased size of the coefficient $b_1$ as searches 2, 3 and 4 are accomplished. Since virtually all movements were away from the university, the negative $b_1$ sign suggests that search adjustment tended to be greater the nearer to the centre that the early searches were made. This is to be expected given the adjustment process traced out in figure 3.3 above. For subsequent searches, $b_1$ retains a significant negative value but falls in absolute size. The influence of search costs can be attributed to the variables G and L. The coefficient signs are in accord with a priori expectations, though not always significant. The degree of general excess demand tends to have a relatively small, though on the whole insignificant effect on Ai. Surprisingly income level does not have a significant effect on distance adjustment though this may reflect the relatively uniform distribution of student income.

The results for experienced students display similarities and contrasts (Table 3.2). Income is once again insignificant, group size effects have a similar size, sign and significance and local origin effects have the same sign as for newcomers but have a less pronounced effect. The generalised excess demand proxy has a more marked, significant effect on Ai than for newcomers. However, the behaviour of $b_1$, the distance adjustment coefficient is quite different. For experienced students, $b_1$ is small in size variable in sign and statistically insignificant for the first five adjustments. In the final adjustment $b_1$ has a small but significant negative sign. Such a pattern tends to suggest that distance adjustment is not the sequential process it is for newcomers; this may be attributable to the differing spatial expectations of experienced students. Nevertheless, outward search adjustment is triggered by temporal increases in excess demand and differences in search costs.

The pattern of results observed in table 3.2 does tend to suggest that adjustment occurs within the local housing market in a fashion consistent with our a priori model. However, this model was designed specifically to be compatible with empirically observed generalisations concerning the expectations and search behaviour of rental tenants. In order to investigate the search behaviour of potential purchaser entrants a different approach must be adopted.

## SEARCH AND OUTCOMES IN THE OWNER OCCUPIED SECTOR

The acquisition of privately owned housing differs in character from the acquisition of rental accommodation (in the context of the City of Glasgow), in three important ways. Firstly, there is substantial variation in property prices/quality characteristics and hence search was intended not only to establish the existence of trading opportunities, but also to provide a comparative valuation of price/quality characteristics. Secondly, unlike the private rental sector in Glasgow, where letting was generally conducted on a "first-come-first-served" basis, house purchase was subject to a closed bidding system in which prospective buyers submitted offers to the seller's agent, the highest offer being accepted. Thirdly, as house purchase cannot for most households be funded solely from income, prospective purchasers must search not only for desirable properties, but also housing finance.

The process of purchase is therefore in many ways much more complex than that of acquiring rental accommodation. Though our problem remains the same in principle - i.e. the influence of the search process upon outcomes - the analytical framework adopted reflects the additional considerations outlined above.

## A Model of Search Experience and Outcomes

The approach we take here is that of a formal examination of the link between search experience and the consumption/expenditure of potential purchasers. To isolate the relevant influences we ignore the household's choice of search procedures, and treat the choice problem purely in terms of current ex-ante unit price offer and housing service consumption given the household's search history. As noted earlier, a closed bidding system is the operational means by which a successful purchase is determined, hence unit price offers made by the household are treated as a choice variable. To render the influence of the search process operational, we assume that all opportunity costs incurred during this process can be mapped by the household into a monetary value, which we label frictional costs.

The determinants of frictional costs are given by the following general functional form.

$$F = f(\bar{S}_L,\ P_B,\ \bar{B}_F,\ I(\bar{S}_L)) \tag{29}$$

where: $\dfrac{\partial F}{\partial \bar{S}_i} \lesseqgtr 0,\quad \dfrac{\partial F}{\partial P_B} < 0,\quad \dfrac{\partial F}{\partial \bar{B}_F} > 0,\quad \dfrac{\partial F}{\partial \bar{I}} > 0,\quad \dfrac{d\bar{I}}{d\bar{S}_L} < 0.$

Where F is frictional costs, $\bar{S}_L$ is length of search, $P_B$ is price offer per unit of housing services, $\bar{B}_F$ is the number of failed bids, $\bar{I}$ is an index of the degree to which market information is absent or obsolete. Given our assumptions concerning past search experience, $\bar{S}_L$, $\bar{B}_F$, $\bar{I}$ and $\bar{E}$ are treated as exogeneously determined.

Planned unit price offer enters as an argument, because in a local housing market where an instantaneous market clearing mechanism does not exist there will in general be spatial and temporal variation in the level of excess demand. The formulation of unit price offers for a given level of housing services acquires significance, because it determines the number of vacancies established by past search activities which satisfy the household's perceived opportunity set. The lower the planned unit price offer the fewer will be the number of acceptable vacancies and the greater will be risk of failure, which can be presumed a source of concern to the household. Under such circumstances the formulation of a higher unit price offer can be expected to reduce the burden of risk.

All other variables in (29) depict the parameters of past search experience. $\bar{S}_L$ represents time devoted to searching, which inevitably involves the incurrence of foregone opportunities. There are benefits in terms of mitigating the absence or obsolescence of market information ($\bar{I}$). $\bar{I}$ is a parameter of relevance because the possession of market information facilitates the ability of the household in displaying "alertness" to trading opportunities, and to discriminate between those of a fruitful or unfruitful nature. Finally, the inclusion of the past number of failed bids reflects the costs associated with the use made of market institutions in making a bid.

(Until recently conveyancing in Scotland could not be undertaken by the individual.)

When utility functions are non-uniform across households and imperfect information characterises the trading process, submarkets can arise as defined in terms of a spatial dispersion of market prices. Given imperfect information households will be uncertain about the dispersion which is established at any point in time. It is against this background that trading plans are formulated and as such it may be considered appropriate to explicitly introduce the individual's search history as a direct influence upon the formulation of planned choices. We posit the following general functional forms for $P_B$ and $H_S$ (housing services).

$$P_B = P_B\,(\bar{B}_F,\ F,\ H_S) \qquad (30)$$

where: $\dfrac{\partial P_B}{\partial \bar{B}_F} > 0,\ \dfrac{\partial P_B}{\partial F} > 0,\ \dfrac{\partial P_B}{\partial H_S} \gtreqless 0.$

$$H_S = H_S\,(P_B),\ \frac{dH_S}{dP_B} < 0. \qquad (31)$$

Functions 29, 30, and 31 together possess two particular characteristics. Firstly, simultaneity is introduced into the relationship between F and $P_B$ and $P_B$ and $H_S$. Secondly, the introduction of determinants directly defining a mapping rule for choice variables is unconventional. These features can be rationalised by recognising that in the complex, uncertain environment underlying this model, households will employ trading procedures whose basis is past experience. Furthermore, this formulation has the plausible feature of allowing price bids and planned housing service consumption the dual role of conforming with the influences exerted by past search, whilst simultaneously being used to flexibly alter the flow of funds between frictional costs and expenditure on other goods and services.

In view of the above discussion both $\bar{B}_F$ and F are included in (30). $\bar{B}_F$ enters separately as an argument in recognition of the price specific information which they yield. For a given level of $H_S$ the failure of a bid indicates that upward revision in the future is required for successful purchase. The relevance of F is its presence as a drain upon current income and hence upon the ability of a household to remain active in the market. Furthermore they represent at any stage prior to successful purchase a "sunk" cost for which the household has nothing to show. For both these reasons F will exert an inflationary influence upon unit price offers. Lastly in this context, experience combined with the household's learning faculties we suggest will lead to the perception of a market required adjustment between $P_B$ and $H_S$. We postulate that households draw upon their experience to construct expectations concerning the unit price the market is willing to accept for a level of housing services. Additionally, the quantity of housing services which the household perceives as feasible with regard to its opportunity set will vary negatively in relation to the unit price offer.

In order to deduce implications concerning planned choices the framework of constrained utility maximisation is adopted. We assume for such purposes the existance of a "well-behaved" utility function whose arguments are housing services and income net of total intended house purchase costs, the latter being defined as equal to expenditure on all other goods and services. Using the notation U for total utility, E for expenditure on all other goods and services and Y for income, we can define

$$U = U(H_S, E) \quad \frac{\partial U}{\partial H_S} > 0, \quad \frac{\partial U}{\partial E} > 0. \tag{32}$$

$$E = \bar{Y} - (P_B H_S + F). \tag{33}$$

The appropriate formulation for the household budget constraint is then

$$\bar{Y} = P_B H_S + F + E. \tag{34}$$

Assuming utility maximising behaviour by households, $H_S$ and $P_B$ will be chosen in such a way as to satisfy those conditions at the margin for a maximum. The problem posed is then

$$\text{Max. } U(H_S, E) \quad \text{subject to} \quad \bar{Y} = P_B H_S + F + E. \tag{35}$$

Using the Lagrange multiplier method, the augmented objective function to be maximised is

$$L = U(H_S, E) + \lambda(\bar{Y} - P_B H_S - F - E). \tag{34}$$

If all functions are everywhere continuously differentiable, the first order conditions for a maximum are:-

$$\frac{\partial L}{\partial H_S} = \frac{\partial U}{\partial H_S} - \frac{\partial U}{\partial E}\left(P_B + H_S \frac{\partial P_B}{\partial H_S} + \frac{\partial F}{\partial P_B}\frac{\partial B}{\partial H_S}\right) = 0 \tag{37}$$

$$\frac{\partial L}{\partial P_B} = \frac{\partial U}{\partial H_S}\frac{dH_S}{dP_B} - \frac{\partial U}{\partial E}\left(H_S + P_B \frac{dH_S}{dP_B} + \frac{\partial F}{\partial P_B}\right) = 0 \tag{38}$$

$$\frac{\partial L}{\partial \lambda} = \bar{Y} - P_B H_S - F - E = 0 \tag{39}$$

To simplify the presentation, let

$$\left(P_B + H_S \frac{\partial P_B}{\partial H_S} + \frac{\partial F}{\partial P_B} \frac{\partial P_B}{\partial H_S}\right) = \delta \tag{40}$$

$$\left(H_S + P_B \frac{dH_S}{dP_B} + \frac{\partial F}{\partial P_B}\right) = \delta \tag{41}$$

Dividing (37) by (38) and rearranging yields the condition

$$\frac{\frac{\partial U}{\partial H_S}}{\delta} = \frac{\frac{\partial U}{\partial H_S} \cdot \frac{dH_S}{dP_B}}{\delta} \tag{42}$$

which can be interpreted as indicating that planned housing service consumption and price per unit offers are adapted to levels such that, the marginal utility per pound incurred on expenditure generated by adjustments to housing service consumption, are exactly balanced by the marginal utility loss per pound incurred on expenditures attributable to unit price offer adjustments. This condition will be satisfied given the values taken by the exogeneous variables $\bar{S}_L$, $\bar{I}$, $\bar{B}_F$ and $\bar{Y}$. Hence the equilibrium planned housing service consumption $\hat{H}_S$ and price offer per unit of housing services $\hat{P}_B$ can be specified as functions of the exogeneous variables, that is:

$$\hat{H}_S = \hat{H}_S(\bar{S}_L, \bar{I}, \bar{B}_F, \bar{Y}) \tag{43}$$

$$\hat{P}_B = \hat{P}_B(\bar{S}_L, \bar{I}, \bar{B}_F, \bar{Y}) \tag{44}$$

Since our analysis has not assumed market equilibrium there is no presumption that these planned magnitudes will be successfully traded. If frictional costs are sufficiently significant and utility functions are not strictly concave then it is conceivable for some households to find themselves during the entry process satisfying marginal conditions compatible with a corner solution, in which case they withdraw from the entry process. For those households who are successful purchasers, planned and relised expenditures are equated at some point in the entry process.

However, in the form specified the model is incapable of ensuring continuity of the functions $\hat{H}_S$ and $\hat{P}_B$ or of yielding comparative static predictions. A strictly concave utility function does not ensure that the second order conditions are met. The reasons can be clarified by the use of figure 3.4. In figure 3.4 the choice framework can be depicted as follows. Suppose we let $H_S$ take some given value $Ho_s$; the household's unit price offer influences the level of frictional costs and hence the level of E available at $Ho_s$. There is then what could be conceived of as a two stage choice problem. Firstly, what is the most effective division of funds at $Ho_s$

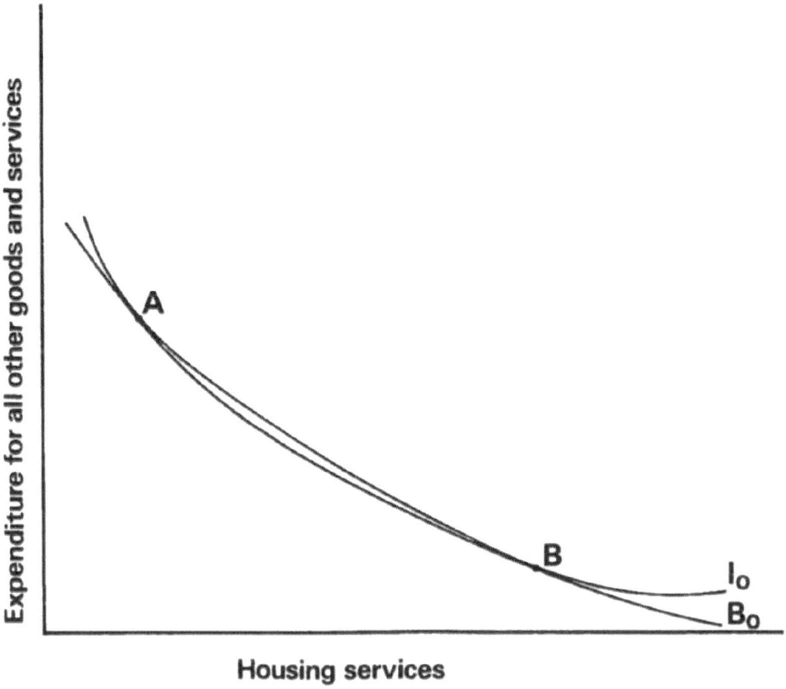


Figure 3.4. Cases of multiple equilibria.

and all other feasible levels of $H_S$. Secondly, which of the alternative optimum divisions of funds maximise utility.

Assume for the purposes of illustration that past experience indicates to the household that the unit price offers which are expected to be successful in the market, decline with increases in $H_S$. If relationships 1 and 3 above are invariant with regard to $H_S$ the budget line becomes convex, as produced by the first stage of the choice problem. Hence the budget line which reflects optimum levels of E for given levels of $H_S$ and $\bar{Y}$, will exhibit a reduced absolute slope with increasing levels of $H_S$. Cases of multiple equilibria as at A and B in figure 3.4 cannot then be ruled out given convexity of indifference curves (I). This creates problems of ambiguity for empirical investigation.

## Empirical Analysis

The data used in this section is derived from interviews undertaken with a sample of 504 first time purchasers in the city of Glasgow in the summer of 1977. The survey material was designed in such a way as to establish the initially held aspirations of prospective purchasing households with regard to housing attribute and expenditure intentions, the difficulties experienced during search activity and the attributes and expenditures incurred on the properties actually purchased. This material is extensive enough to provide empirical proxies for the factors highlighted above, and enables us to examine the adjustments which households made to planned housing expenditure in securing entry into the owner occupied housing sector.

To investigate these adjustments we take as our dependent variable the discrepancy between actual housing expenditure and the housing expenditure aspired to at the onset of search. We label this measure the "premium adjustment", and interpret it as the magnitude of revised plans contained in the successful bid. In line with our earlier analysis we expect its size to be influenced by the household's past search history. However, not only do we face the problems of an absence of unambiguous a priori expectations, we are also unable to disentangle "premium adjustments" brought about by revision to unit price offers and/or planned housing service consumption. An attempt is made later in this section to mitigate this problem.

Our survey material allowed the construction of proxies or direct measures for those exogeneous variables $B_F$, I and $S_L$ which were stressed earlier. In addition, it was possible to obtain a measure of market agent fees ($S_c$) which we have added to our stock of independent variables. Our model ignored the issue of the correct formulation of the budget constraint. Two issues are raised here; firstly, it is commonly alleged that permanent and not current income is the magnitude which households refer to when deciding upon planned levels of housing expenditure. Secondly, as noted in the introduction most households must rely upon access to loan finance and hence the process of purchase involves search not only for a suitable property, but also for funds commensurate with intentions. Thus when initial intentions are not backed by a suitable timing and availability of funds a further factor influencing "premium adjustment" is apparent. To the extent to which lending institutions base their lending rules upon clients' permanent income, these points are taken account of by the inclusion of a credit constraint variable C, the measurement of which is explained in Table 3.3. Further, since loans are commonly "geared" the extent to which the credit constraint binds will depend upon the stock of wealth or savings which households hold and can contribute to the cost of purchase. This ability we suggest to be an additional influence on "premium adjustment" and hence we include a wealth variable W in our estimated equations.

The independent variables are listed and their measurement explained in Table 3.3. The linear specification of the hypothesised relationship is:

$$A = b_0 + b_1 B_F + b_2 I + b_3 Sc + b_4 N + b_5 c + b_6 S_L + u \quad (45)$$

where u is the error term. A number of comments are pertinent on the question of measurement. Firstly, the wealth variable used is the size of deposit in the purchase price and this will be an imperfect

Table 3.3: Regression Results: Housing Expenditure Adjustment

| Dependent Variable | | Coefficients for the Independent Variables (Standardized Beta Coefficients in Parentheses) | | | | | | | | |
|---|---|---|---|---|---|---|---|---|---|---|
| | | Constant | $B_F$ | I | Sc | W | C | $S_L$ | Adjusted $R^2$ | F |
| h(1) | A | 652.3 | 484.8* | -94.4 | -0.62* | 0.42* | -2,100.7* | 27.8** | 0.53 | 28.22* |
| | | | (.17) | (-.02) | (-.21) | (.61) | (-.53) | (-.11) | | |
| h(2) | P | 249.0 | 614.6** | 633.9 | 19.66** | 0.45** | -1,347.1* | 12.3 | 0.57 | 9.09* |
| | | | (.24) | (.18) | (.28) | (.69) | (-.43) | (.09) | | |

* Significance at 0.01
** Significance at 0.05

Where A = Actual housing expenditure minus aspired to housing expenditure at the onset of search (n = 144).
P = Unit price premium.
$B_F$ = Number of failed bids.
I = Information constraint, I = 1 for those who felt that lack of information on available housing and the spatial dimensions of the market inhibited search, I = 0 otherwise.
Sc = Market agent fees and is equal to the sum of the lawyer and surveyor fees.
W = Wealth proxy, and is the size of deposit in purchase price.
C = Credit constraint, C = 1 if mortgage amount advised less than aspired to housing expenditure. C = 0 otherwise.
$S_L$ = Length of search

proxy to the extent to which households differ in the propensity with which they devote savings or wealth to housing expenditure. Secondly, the length of search variable is a little unsatisfactory because it makes no adjustment for differing intensities. Both comments suggest caution with regard to interpretation of the estimated coefficients.

The results reported in Table 3.3 are based upon the estimation of our linear specification by ordinary least squares estimation techniques. The linear specification gives a reasonably good fit, given that we are dealing with cross section data. Though expected signs are a priori indeterminate a significant pattern emerges; five of the six independent variables are significant, four at the 0.01 level. The wealth and credit constraint variables have coefficients of the opposite sign and are the strongest explanatory variables. The number of failed bids and length of search variables are significantly positive, while market agent fees and the information constraint are negative, though the latter is insignificant.

Our formal model helps clarify the problems involved in interpreting the nature of the "premium adjustment", which these estimates imply. The variable $B_F$ is indicative of these problems. As a component of past search experience and contributor to frictional costs, $B_F$ can be viewed as an inflationary influence upon unit price offers; however, the market agent fees which accompany failed bids will have a restrictive influence upon the budget constraint and hence can be expected to produce downward revision in planned housing expenditure. Since the estimated coefficient sign is positive, the former influence would appear to dominate. This is to be expected as our regression model includes market agent fees separately and may then capture the influence of those financial costs sustained due to failed bids. The negative coefficient estimate which accompanies the market agent fees variable substantiates this interpretation. The two remaining components of frictional costs, $S_L$ and I, have similarly complicated channels of influence. The significant positive coefficient upon $S_L$ suggests that the opportunity costs of search are an inflationary influence upon price offers which outweigh the restrictive impact upon financial constraints.

The W and C variables are somewhat easier to interpret. The existence of a binding credit constraint can be expected to deflate planned unit price offers and produce downward adjustment in planned housing service consumption, as households perceive and experience the additional limitation upon their feasible set. The strength and sign of the estimated coefficient is in accord with these anticipations. The strong positive sign attached to the coefficient on the wealth variable, implies that size of household's wealth holding will be an important factor in mitigating the limitations which frictional costs and the credit constraint pose for the household's feasible set.

As noted earlier, the existence of both quantity and price adjustments as components of the dependent variable complicates interpretation. As an attempt to clarify the empirical analysis the regression model was estimated for that subgroup of the sample who satisfied their aspirations with regard to the major attributes of housing (Table 3.3). Hence the subgroup is comprised of those households who made no change to quantity aspirations during search activity, and if any adjustments are made they will be with regard to the unit price offer and/or expenditure on all other goods. The dependent variable can therefore be interpreted in this context, as the "unit price premium" paid by households in order to satisfy quantity aspirations.

The fit is again reasonably good, but statistical significance suffers as is to be expected given the fewer degrees of freedom. In particular $S_L$ becomes insignificant, though I now just fails to be significant at the 0.05 level with a positive coefficient as opposed to the negative coefficient reported in the full sample of households. The variables $B_F$, W, C and $S_c$ retain their statistical significance with $B_F$ and W stronger determinants than they are with regard to the full sample of households. The S coefficient changes sign and also exerts a stronger influence than in the regression model applied to the full sample.

Thus despite satisfaction of quantity aspirations this subgroup made important adjustments to their unit price offers during search. Once again, the credit constraint variable clearly represents an important limitation here. However, size of wealth holdings, the number of failed bids and search costs all appear to be significant factors exerting a positive influence on unit price adjustments. For this group of households quantity aspirations would appear to be met by:

(1) Discounting market agent fees into price offers.
(2) Using the information contained in failed bid attempts to arrive at an increased unit price offer which the market will bear.
(3) Using wealth holdings to facilitate upward price adjustment.

## CONCLUSION

This chapter is based on the premise that uncertainty and imperfect information impute a significance to the process of movement which warrant investigation in the housing market. Conceptual categories were defined to facilitate analysis of this process, and the channels through which past search history could influence planned choices were formally examined with regard to both rental and owner occupied housing. Empirical investigation concentrated upon the adjustments to planned choices which past search history was hypothesised to influence. Household aspirations held at the onset of search activity were used as a reference point from which to measure the adjustments made during the process of trading. Multiple regression techniques were employed to investigate the determinants of these adjustments and to test the hypothesis that frictional elements arising from imperfect information and uncertainty, are an important factor explaining the difficulties faced by households in meeting initially held aspirations.

This approach to housing choice opens up a number of policy issues which have traditionally received little attention. If as a consequence of the search process being sequentially juxtaposed between the decision to consider moving and the actual outcome, its characteristics are an important influence upon the latter, then there are implications for demand orientated housing programs. Such programs generally presume that households in receipt of benefits will be able to negotiate successfully for themselves. As the results of this paper imply, the complexity of the search process may militate against this presumption. In particular, the time, effort and monetary costs involved in the process of purchase, may debilitate the effectiveness of mobility as a means of improving an individual household's housing standards.

## NOTES

1. Formulation of the problem in this way reflects the earlier observation that vacancies exhibit little variation in price/quality characteristics. Hence, adjustments in housing related expenditure can be primarily attributed to search costs. To simplify the analysis, the model's specification implicitly assumes that the individual is unwilling to trade-off consumption of other goods in order to sustain search costs above the maximum intended.

2. If expected search costs given initial search location are below the maximum intended and if the constraint is maintained as a strict equality, then the adjustment process is of the same character but in the opposite direction.

## REFERENCES

McCarthy, K. (1980). Housing Search and Consumption Adjustment Santa Monica, California: Rand Corporation, P-6473.

Maclennan, D. (1977). "Some thoughts on the nature and purpose of urban house price studies", Urban Studies, 14, 59-71.

----- (1978). "The 1974 Rent Act: some short-run supply effects", Economic Journal, 88, 331-340.

----- 1979(a). "Information and adjustment in a local housing market", Applied Economics, 11, 255-270.

----- 1979(b). "Information networks in a local housing market", Scottish Journal of Political Economy, 26, 73-88.

# 4

# RACIAL DIFFERENCES IN THE SEARCH FOR HOUSING

**Francis J. Cronin**

Recently, researchers have applied models of optimal search behavior to explain various adverse market conditions (e.g. price differentials) facing households of particular racial minority groups (Masson 1973; Courant 1978).[1] Essentially, the researchers introduce seller prejudice, or the perception of seller prejudice on the part of buyers, into models of buyer or renter search. When applied to the housing market, the models imply that minority households will, ceteris paribus, be less likely to search in general, and will be more likely to search in neighborhoods with higher proportions of minority households than they otherwise would. The first result suggests that minority households are more likely to accept a less optimal dwelling unit; the second result suggests that the current distribution of neighborhood racial composition is likely to persist. More importantly, the models imply that the adverse market conditions are not likely to diminish as a result of natural market forces, given the rational response of households to the prejudice confronting them.

Given the above implications, it is surprising that so little research has empirically investigated racial differences in housing search.[2] Two obstacles probably account for the dearth of research on housing search in general, and racial differences in particular. These are, the difficulty of specifying optimal search models in an applied context,[3] and the lack of data on housing search behavior with which to estimate search models.

This chapter reports the results of estimating such differences in housing search between nonminority and minority households within the context of optimal search models.[4] The second section presents a brief outline of the theoretical framework and the development of the testable hypotheses. In particular, the household's gain from relocation is derived. The third section presents the results of testing some of these implications using a longitudinal panel of households. Differences in the sources of information and modes of transportation employed by nonminority and by minority households in housing search are examined. Differences in household search behavior (i.e., number of days spent searching, number of units searched, number of units phoned about, and number of neighborhoods searched, average and extreme radius

DOI: 10.4324/9781003182085-5

of search) between nonminority and minority households are estimated. Differences in the characteristics of neighborhoods searched in and moved to by nonminority and minority are also examined. Finally, differences between the characteristics of the neighborhoods of origin, search, and destination are examined for nonminority and minority households.

## THE THEORETICAL FRAMEWORK

### The Gains From Moving

We assume each household to be a rational utility maximizer. Therefore, the household will relocate if, in so doing, the household expects to be made better off.

Consider a household in two-commodity space -- housing services, H, and nonhousing goods and services, X.[5] The household's preference ordering is assumed to be representable by an ordinal utility function

$$U = u(H, X) \tag{1}$$

which meets the basic axioms of preference relations.[6] The household maximizes (1) subject to a budget constraint

$$Y = PhH + PxX \tag{2}$$

where Y is the income of the household, Ph is the price per unit of housing service, and Px is the price per unit of X. The resulting set of first order conditions

$$\frac{Uh}{Ux} = \frac{Ph}{Px} \tag{3}$$

implies an optimal commodity bundle, (H*, X*), given by

$$H^* = g(Y, Ph, Px) \tag{4a}$$

$$X^* = (Y - PhH^*)/Px \tag{4b}$$

Equations (1) through (4) permit derivation of the household's potential gain from relocation. This gain, the equivalent consumer's surplus (Hicks, 1956), is the amount of additional income (Ye - Y) that would make the household as well-off with its current consumption of housing as it would be if it were to relocate and consume its optimal quantity of housing services. The equivalent consumer's surplus (ECS) is given by

$$u(H, (Ye - PhH)/Px) = u(H^*, (Y - PhH^*)/Px) \tag{5}$$

$$ECS = Ye - Y \quad (6)$$

It is significant to note that the equivalent consumer's surplus is a cardinal measure which can be estimated from household data.

The equivalent consumer's surplus is depicted graphically in Figure 4.1. In time period 0 the household maximizes its utility at C on indifference curve Uo, given its budget constraint AB, and consumes Ho* of housing and Xo* of other goods and services. If in time period 1 the household's income increases to A'B', the utility maximizing position would be at D on indifference curve $U_1$. Optimal housing consumption would increase to $H_1$*. The equivalent consumer's surplus is the grant of nonhousing goods and services (in

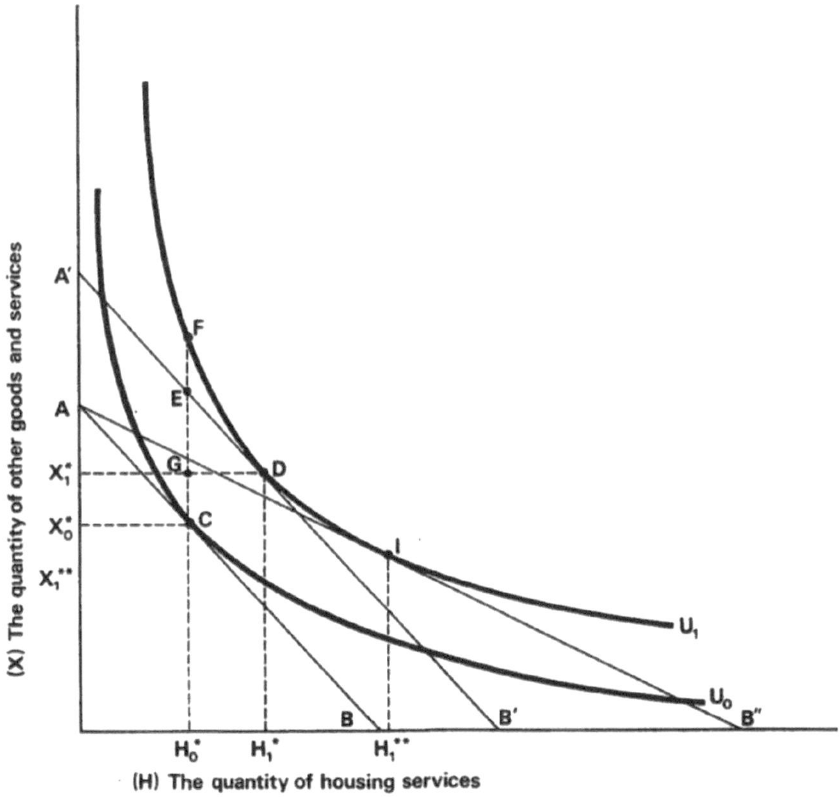


Figure 4.1. The household's gain from relocation.

income terms) that places the household on $U_1$, holding its housing consumption at Ho*. In Figure 4.1, this grant is CF-CE.

The equivalent consumer's surplus is a general measure of gain; inducements to relocate based on changes in either equation (1) or (2) are easily handled. For example, the gain to the household from a change in the relative price of housing can be handled just as we did for a change in income. In this case, if the new relative price of housing is given by AB" in Figure 4.1, the equivalent consumer's surplus of relocating to a unit providing $H_1$** of housing services would be CF. Likewise, changes in the household's "taste" for housing (i.e., the parameters of the indifference map) can be dealt with.

## The Optimal Level Of Housing Search[7]

We assume that households know the distribution of available units in terms of the level of H they provide. Knowing their utility functions, knowing the distribution of available units, and assuming that households would not consider a unit less optimal than its current unit, households are able to compute a distribution of gains (i.e., an ECS for each available unit). Denote this distribution f(G). The distribution f(G) meets the standard condition

$$\int_{Gmin}^{Gmax} f(G)dG = 1 \tag{7}$$

where Gmin is the lowest gain considered (i.e., the gain associated with the household's current unit, (0) and Gmax is the gain associated with the household's utility maximizing unit. The cumulative distribution of f(G) is denoted F(G).

At any stage of the housing search process, the value of the household's gain from one additional search is

$$G = \max \begin{cases} Go \\ -SC + (1-F(Go))E(G \mid G \geq Go) + GoF(Go) \end{cases} \tag{8}$$

Go is the gain associated with the best unit found thus far.[8] The cost of searching one unit is denoted SC. The second line of the right-hand side of (8) is the expected gain after one more search. A rational household stops searching when the first line of the right-hand side of (8) is greater than or equal to the second line. That is, when

$$SC = \int_{Go}^{Gmax} (G-Go)f(G)dG \tag{9}$$

At this point, the expected gain from an additional search is just equal to the costs of search.

Equation (9), for a given f(G), defines an optimal gain (G*) at which to cease search. The value of $G_i^*$ for the household depends on the values of $Gmax_i$ (we already know Gmin), $SC_i$, and the form of $f_i(G_i)$.

Assume now that minorities perceive (either correctly or incorrectly) that there is some nonzero probability ($\alpha$ j) that nonminority landlords in neighborhood j will be unwilling to rent to them at current market prices. The probability that a minority household searching in neighborhood j will find a landlord averse to it, is $\alpha$j times the fraction of landlords in neighborhood j who are nonminorities ($\beta$j), or $\alpha j \beta j = \delta j$. The effect of landlord aversion is to set $\delta$j of the density f(G) equal to zero. This reduces the expected gain from search in neighborhood j, or increases the cost. On average, a minority household searching in neighborhood j will have to look at $1/(1-\delta j)$ units for every unit it has the option of renting.

This perception on the part of minority households results, ceteris paribus, in a ranking of neighborhoods to be searched starting with the neighborhood with the lowest value of $\delta$. Furthermore, the optimal amount of search is reduced since

$$SC/(1-\delta) > SC \tag{10}$$

for $0 < \delta < 1$. This increase in search cost reduces the level of search at which an expected incremental gain equals the now higher level of search cost.

## THE EMPIRICAL ANALYSIS

Fortunately, many of the theoretical implications discussed above can be examined empirically with data now becoming available from the Housing Allowance Demand Experiment (HADE). This experiment, conducted for the Department of Housing and Urban Development during the mid-1970's was designed to test the responses of low-income households to income and price subsidies for housing.[9] Longitudinal data on the search and relocation behavior of households from the HADE sites of Allegheny County (Pittsburgh) and Maricopa County (Phoenix) are employed in the analysis below.

### Housing Search Behavior

As discussed above, the optimal amount of search given by equation (9) depends on the household's values of Gmax and SC and on the form of f(G).

The derivation of Gmax has already been discussed in the previous section. To estimate Gmax, the utility function must be specified in an estimable form; not every explicit utility function implies a closed-form solution for the equivalent consumer's surplus.[10]

Nor does every utility function imply unrestricted price and income elasticities of demand.[11] Due to its closed-form solution, unrestricted price and income elasticities of demand, and advantage in estimation, the Stone-Geary form of the house hold's indifference map has been selected for the empirical analysis. Cronin (1980a) provides more information on this choice.[12]

We assume that the household's cost of search in the short-run varies with the sources of information and modes of transportation employed in search, with the household's opportunity cost of time, and with changes in the household's behavior due to the expectation of discrimination. Tables 4.1 and 4.2 present data for the second six-month period of the experiment on the sources of information employed by searchers, the source of information used by movers to find a unit, and the effectiveness of each information source in Pittsburgh and Phoenix, respectively.

Note the generally high rate of usage for many of the sources, particularly Newspapers, Vacancy Signs, and Friends or Relatives. The pattern of sources used in each site is similar. However, households in Pittsburgh tend to use more sources than those in Phoenix and minority households in Pittsburgh employ the services of a Real Estate Agency at a much higher rate than do minority households in Phoenix.[13] In each site, two statistically significant differences are apparent between the sources used by nonminority and minority households with the only consistent difference being minority households' higher use of Vacancy Signs.[14]

Examining the data on the source leading to a unit and the effectiveness of each source (i.e., the percentage of searchers using a source that found their unit via that source), we again see a similar pattern between the sites with Friends or Relatives being the source most frequently cited as that leading to a unit. Friends or Relatives is also generally the most effective source. The only statistically significant difference between nonminority and minority households is in the success rate and effectiveness of Newspapers in Phoenix.

The usual mode of transportation employed during search in the second six-month period of the HADE is displayed in Tables 4.3 and 4.4. Note that minority households compared with nonminority households are significantly less likely to use their own car. A similar difference is apparent when households with access to a friend's or relative's car are accounted for. Minority households in Pittsburgh are more likely to rely on a taxi or jitney, public transportation or walking. Minority households appear, particularly in Pittsburgh, to have a higher marginal cost of search when compared with nonminority households.

The household's perception of f(G) is assumed to vary with (1) the extent of the household's recent mobility (i.e., number of moves in the previous three-year period, (2) the education of the household head, (3) the age of the household head, and (4) the size of the household. The estimated model also includes the race of the household head. As specified, it is not possible to infer whether the

Table 4.1: Sources of Information and Their Effectiveness in the Second Six-Month Period of the Experiment in Pittsburgh

| | Sources Used | | | Source Leading to Unit | | | Effectiveness[b] | | |
|---|---|---|---|---|---|---|---|---|---|
| Source of Information | Non-Minority | Minority | Chi-Square[a] | Non-Minority | Minority | Chi-Square[a] | Non-Minority | Minority | Chi-Square[a] |
| Newspapers | 67.5% | 62.6% | .50 | 23.5% | 15.0% | .27 | 11.88% | 5.26% | 1.36 |
| Real Estate Agency | 43.9 | 56.0 | 3.43 | 8.6 | 15.0 | .19 | 6.54 | 5.88 | .02 |
| Neighborhood Bulletin Board | 14.3 | 15.4 | .00 | - | - | - | - | - | - |
| Vacancy Signs in Buildings | 30.8 | 49.5 | 9.13** | 1.2 | 5.0 | .03 | 1.33 | 2.22 | .15 |
| Friends or Relatives | 66.2 | 59.3 | 1.08 | 60.5 | 55.0 | .04 | 31.21 | 20.37 | 1.82 |
| Other | 7.6 | 14.3 | .00 | 6.1 | 10.0 | .01 | 27.78 | 15.38 | .14 |
| Sample Size | (237) | (91) | | (81) | (20) | | | | |

*Sample:* Households present for the second six-month period of the experiment, paying full-market rents in nonsubsidized rental housing, and searching for alternative housing.

[a] Comparing the source of information employed by households in each racial group.

[b] Effectiveness is defined as the percentage of searchers using a source that found their new dwelling via that source.

*Statistic significant at .05 level.
**Statistic significant at .01 level.

Table 4.2: Sources of Information and Their Effectiveness in the Second Six-Month Period of the Experiment in Phoenix

| Source of Information | Sources Used | | | Source Leading to Unit | | | Effectiveness[b] | | |
|---|---|---|---|---|---|---|---|---|---|
| | Non-Minority | Minority | Chi-Square[a] | Non-Minority | Minority | Chi-Square[a] | Non-Minority | Minority | Chi-Square[a] |
| Newspapers | 61.0% | 40.0% | 10.85** | 41.7% | 12.1% | 14.99** | 42.55% | 17.39% | 7.60** |
| Real Estate Agency | 20.8 | 13.0 | 2.22 | 1.0 | 3.0 | .11 | 3.13 | 13.33 | .48 |
| Neighborhood Bulletin Board | 9.7 | 15.7 | 1.62 | 1.0 | 0.0 | .04 | 6.67 | 0.00 | .01 |
| Vacancy Signs in Buildings | 50.0 | 62.6 | 3.35 | 15.6 | 27.3 | 2.59 | 19.23 | 25.0 | .38 |
| Friends or Relatives | 46.8 | 56.5 | 2.14 | 34.4 | 48.5 | 2.68 | 45.83 | 59.26 | .05 |
| Other | 9.6 | 8.4 | 0.01 | 6.3 | 9.1 | .14 | 40.91 | 62.17 | .33 |
| Sample Size | (154) | (115) | | (96) | (66) | | | | |

*Sample:* Households present for the second six-month period of the experiment, paying full-market rents in nonsubsidized rental housing, and searching for alternative housing.

[a] Comparing the source of information employed by households in each racial group.

[b] Effectiveness is defined as the percentage of searchers using a source that found their new dwelling via that source.

*Statistic significant at .05 level.

**Statistic significant at .01 level.

Table 4.3: Mode of Transportation Employed for Searching During the Second Six-Month Period in Pittsburgh

| Mode | Nonminority | Minority | Chi-square[a] |
|---|---|---|---|
| Own Car Usually Used | 40.9 | 18.7 | 13.39** |
| Excluding Households Usually Using a Friend's or Relative's Car | 35.4 | 17.6 | 9.07** |
| Friend or Relative's Car Usually Used | 32.9 | 37.4 | 0.53 |
| Excluding Households Usually Using Own Car | 27.4 | 36.3 | 2.04 |
| Taxi or Jitney Usually Used | 2.95 | 15.38 | 14.94** |
| Excluding Households Usually Using Own Car | 1.69 | 9.89 | 9.57** |
| Excluding Households Usually Using Own, Friend's or Relative's Car | 1.27 | 5.49 | 3.32 |
| Public Transportation Usually Used | 29.54 | 56.04 | 18.72** |
| Excluding Households Usually Using Own Car | 13.08 | 21.98 | 3.32 |
| Excluding Households Usually Using Own, Friend's or Relative's Car | 13.92 | 28.57 | 8.59** |
| Walk Usually Used | 49.79 | 71.43 | 11.62** |
| Excluding Households Usually Using Own Car | 16.46 | 24.18 | 2.10 |
| Excluding Households Usually Using Own, Friend's or Relative's Car | 24.05 | 37.36 | 5.17* |
| Other Usually Used | 0.00 | 0.40 | 0.25 |
| Sample Size | (237) | (91) | |

*Sample:* Households present for the second six-month period of the experiment, paying full-market rents in non-subsidized rental housing, and searching for alternative housing.

[a]Chi-square statistic tests the difference in the mode of transportation usually employed between racial groups.

*Statistic significant at .05 level.

**Statistic significant at .01 level.

Table 4.4: Mode of Transportation Employed for Searching During the Second Six-Month Period in Phoenix

| Mode | Nonminority | Minority | Chi-Square[a] |
|---|---|---|---|
| Own Car Usually Used | 73.38 | 62.61 | 3.07 |
| Excluding Households Usually Using a Friend's or Relative's Car | 70.78 | 54.78 | 6.64** |
| Friend or Relative's Car Usually Used | 16.23 | 34.78 | 11.37** |
| Excluding Households Usually Using Own Car | 13.64 | 26.96 | 17.97** |
| Taxi or Jitney Usually Used | 0.00 | 2.61 | 2.04 |
| Excluding Households Usually Using Own Car | 0.00 | 2.61 | 2.04 |
| Excluding Households Usually Using Own, Friend's or Relative's Car | 0.00 | 0.87 | 0.02 |
| Public Transportation Usually Used | 3.90 | 4.35 | 0.02 |
| Excluding Households Usually Using Own Car | 2.60 | 3.48 | 0.00 |
| Excluding Households Usually Using Own, Friend's or Relative's Car | 1.95 | 0.87 | 0.05 |
| Walk Usually Used | 23.38 | 31.30 | 1.73 |
| Excluding Households Usually Using Own Car | 12.99 | 23.48 | 4.32* |
| Excluding Households Usually Using Own, Friend's or Relative's Car | 9.09 | 7.83 | 0.02 |
| Other Usually Used | 3.90 | 1.74 | 0.45 |
| Sample Size | (154) | (115) | |

*Sample:* Households present for the second six-month period of the experiment, paying full-market rents in non-subsidized rental housing, and searching for alternative housing.

[a]Chi-square statistic tests the difference in the mode of transportation usually employed between racial groups.

*Statistic significant at .05 level.

**Statistic significant at .01 level.

race variable is acting as a determinant of f(G) or of SC, although as mentioned, a variable representing whether a household avoided certain neighborhoods due to the expectation of discrimination is included in the model and should act as an argument of SC.

We now estimate the implied relationship given by equation (9). Six dimensions of search behavior are examined: the number of days spent searching, the number of neighborhoods searched, the number of dwelling units searched, the number of dwelling units phoned, the average radius of search, and the extreme radius of search.[15]

In general, the regressions performed very well.[16] Significant positive relationships (negative for days searched) were generally found between the measures of search and the gain (the equivalent consumer's surplus) and between the measures of search and less expensive sources of information and modes of transportation.[17] All of the estimated regressions were significant at a .05 level or higher and the proportion of variance explained generally ranged between .10 and .30.

Results of racial differences in the estimated regressions are reported in Table 4.5 for searchers and movers separately by site. In general, the theoretical implications of racial differences in the optimal amount of housing search developed in the second section are strongly supported by the regression results.[18] In Pittsburgh, significant racial differences are apparent in all measures of search behavior -- minority households, on average, search in .68 fewer neighborhoods and 1.52 fewer units, phone regarding 5.4 fewer units, and search .59 fewer miles on average and .92 fewer miles in the extreme. In addition, when minority households do move, their search time exceeds that of nonminority households by 83 days. In Phoenix, significant racial differences are apparent in four measures of search behavior -- minority households, on average, search in 2.16 fewer units, and search 1.36 fewer miles on average and in the extreme. As in Pittsburgh, when minority households in Phoenix do move, their search time exceeds that of nonminority households, in this case by 31 days.

When we examine the behavior of movers only, we see that no racial differences in search behavior are apparent for households in Pittsburgh. In Phoenix, however, differences are found in the number of dwelling units searched and phoned about and in the average and extreme radius of search. In fact, for all but the last measure, the differences between nonminority and minority households are larger for movers only than for all searchers.

## The Characteristics Of Neighborhoods Searched In And Moved To

Characteristics of all neighborhoods searched in during the last 18 months of the experiment (for households searching in five or fewer neighborhoods)[19] as well as the characteristics of neighborhoods moved to at any time during the 24 months of the experiment have been obtained from Fourth-Count census data. Information on the census tracts comprising each neighborhood have been weighted by

Table 4.5: Regression Results on Racial Differences in Search Behavior for All Searchers and for Movers Only By Site During the Second Six-Month Period of the Experiment

| Number of Days Searched (Movers & Households about to Move) | Number of Neighborhoods Searched | Number of Dwelling Units Searched | Number of Dwelling Units Phoned | Average Radius of Search | Extreme Radius of Search |
|---|---|---|---|---|---|
| | | Searchers in Pittsburgh | | | |
| -32.910* | .678** | 1.523* | 5.411** | .594 | .920* |
| (1.78) | (2.42) | (1.70) | (2.67) | (1.63) | (1.89) |
| | | Movers in Pittsburgh | | | |
| - | .287 | -.473 | 5.081 | 0.265 | .397 |
| | (0.48) | (0.23) | (0.94) | (0.03) | (0.41) |
| | | Searchers in Phoenix | | | |
| -30.940* | .079 | 2.156* | 1.368 | 1.361** | 1.358* |
| (1.77) | (0.28) | (2.35) | (0.88) | (2.69) | (2.32) |
| | | Movers in Phoenix | | | |
| - | .064 | 2.794** | 3.424* | 1.384* | 1.190 |
| | (0.20) | (2.61) | (2.03) | (2.00) | (1.54) |

*Sample:* Searchers include all households searching in five or fewer neighborhoods during the second six-month period of the experiment; movers include all households moving during the second six-month period of the experiment. Both samples include only those households paying full-market rents in nonsubsidized rental housing. The race variable is coded 1 for a nonminority and 0 for a minority household. The number in parentheses is the t-statistic.

*t Statistic significant at .05 level.

**t Statistic significant at .01 level.

tract population to form neighborhood data for various characteristics.

The distribution of each characteristic for searchers and for movers has been derived for both nonminority and minority households in each site. Summary descriptors of these distributions along with tests of significance comparing the distribution of each variable by nonminority versus minority status are presented in Tables 4.6 and 4.7 for Pittsburgh and Phoenix, respectively. Two nonparametric tests of significance are shown for each variable. The Mann-Whitney U statistic (M-W) tests the null hypothesis of no difference in the central tendency of the variable's distributions between nonminority and minority households. The Kolmogorov-Smirnov statistic (K-S) tests the null hypothesis of no difference in the variables' distributions between nonminority and minority households.

As Table 4.6 indicates, minority and nonminority households differ in 5 of 8 characteristics describing the types of neighborhoods households searched in or moved to. Compared to nonminority households, minority households search in and move to neighborhoods with: a lower mean household income; a higher proportion of households below the poverty line; a higher proportion of black households; a lower proportion of occupied units that are owner-occupied; and, a lower proportion of occupied rental units with complete plumbing facilities, direct access, and complete kitchen and heating facilities. No statistically significant differences are found for mean gross rent, mean contract rent, or the proportion of dwelling units built before 1939.

In Phoenix (Table 4.7), however, we find a statistically significant difference for all of the characteristics of the search and move neighborhoods of minorities compared with nonminority households. In addition to the differences found in Pittsburgh, in Phoenix, minority households compared with nonminority households, search and move to neighborhoods with lower gross and contract rents and a higher proportion of units built before 1939.

So far, we find that minority households differ from nonminority households in: their selection of search strategies (i.e., sources of information and mode of transportation), their search behavior, the types of neighborhoods searched in, and the types of neighborhoods moved to. Given these differences, it would be informative to compare the characteristics of the neighborhood at origin, the neighborhoods searched in, and the neighborhood moved to for nonminority and minority households separately.

Tables 4.8 and 4.9 present a series of three comparisons for nonminority and minority households separately: origin neighborhood versus search neighborhoods, search neighborhoods versus move neighborhood, and origin neighborhood versus move neighborhood. In addition, tests of significance for each of nine neighborhood characteristics are presented. The null hypothesis of no difference in the distributions of each characteristic is based on a two-tailed Wilcoxon matched-pairs signed-ranks Z statistic.

Table 4.6: Characteristics of Neighborhoods Searched in and Moved to for Nonminorities and Minorities in Pittsburgh

| Characteristic | Neighborhoods Searched In | | | Neighborhood Moved To | | |
|---|---|---|---|---|---|---|
| | Nonminority | Minority | Statistics | Nonminority | Minority | Statistics |
| Mean Household Income (in $1,000) | | | | | | |
| Mean | 10.54 | 9.26 | M-W** | 10.17 | 8.87 | M-W** |
| Variance | 6.66 | 2.82 | K-S** | 4.69 | 4.40 | K-S** |
| Percent of Households Below Poverty Line | | | | | | |
| Mean | 8.68 | 15.30 | M-W** | 9.17 | 16.49 | M-W** |
| Variance | 23.14 | 45.70 | K-S** | 25.36 | 89.55 | K-S** |
| Percent of Population that is Hispanic | | | | | | |
| Mean | --- | --- | --- | --- | --- | --- |
| Variance | --- | --- | --- | --- | --- | --- |
| Percent of Population that is Black | | | | | | |
| Mean | 6.27 | 33.26 | M-W** | 6.22 | 38.96 | M-W** |
| Variance | 106.71 | 489.29 | K-S** | 114.75 | 1002.71 | K-S** |
| Mean Gross Rent (in dollars per month) | | | | | | |
| Mean | 100.65 | 101.24 | M-W | 96.55 | 96.07 | M-W |
| Variance | 21.81 | 14.14 | K-S | 53.17 | 47.47 | K-S |
| Mean Contract Rent (in dollars per month) | | | | | | |
| Mean | 83.48 | 85.46 | M-W | 79.21 | 79.52 | M-W |
| Variance | 18.60 | 11.43 | K-S | 62.83 | 55.40 | K-S |

Table 4.6: (continued)

| Characteristic | Neighborhoods Searched In | | | Neighborhood Moved To | | |
|---|---|---|---|---|---|---|
| | Nonminority | Minority | Statistics | Nonminority | Minority | Statistics |
| Percent of Occupied Units that are Owner Occupied | | | | | | |
| Mean | 14.35 | 13.12 | M-W* | 13.77 | 12.60 | M-W* |
| Variance | 20.16 | 14.29 | K-S | 24.50 | 19.30 | K-S* |
| Percent of Occupied Rental Units with all Plumbing Facilities, Direct Access, Complete Kitchen Facilities, and Heating | | | | | | |
| Mean | 91.50 | 89.27 | M-W** | 91.25 | 89.80 | M-W* |
| Variance | 38.32 | 36.12 | K-S** | 44.28 | 55.27 | K-S |
| Percent of Units Built Before 1939 | | | | | | |
| Mean | 68.31 | 67.35 | M-W | 71.23 | 70.54 | M-W |
| Variance | 328.70 | 194.04 | K-S | 465.89 | 300.69 | K-S |
| Sample Size | (338) | (97) | | (279) | (84) | |

*Sample:* The data on search neighborhoods includes households searching in five or fewer neighborhoods during either the second six-month period and/or the second twelve-month period of the experiment; the data on move neighborhoods includes households moving at any time during the experiment. Both samples include only those households paying full-market rents in non-subsidized rental housing.

*Note:* For both the Mann-Whitney U Test (M-W) and for the Kolmogrov-Smirnov Test (K-S), significance is based on a two-tailed test.
*Statistic significant at .05 level.
**Statistic significant at .01 level.

Table 4.7: Characteristics of Neighborhoods Searched in and Moved to for Nonminorities and Minorities in Phoenix

| Characteristic | Neighborhoods Searched In | | | Neighborhood Moved To | | |
|---|---|---|---|---|---|---|
| | Nonminority | Minority | Statistics | Nonminority | Minority | Statistics |
| Mean Household Income (in $1,000) | | | | | | |
| Mean | 10.92 | 8.99 | M-W** | 11.18 | 9.31 | M-W** |
| Variance | 5.81 | 3.25 | K-S** | 5.34 | 2.82 | K-S** |
| Percent of Households Below Poverty Line | | | | | | |
| Mean | 9.65 | 16.65 | M-W** | 9.60 | 15.07 | M-W** |
| Variance | 38.29 | 85.21 | K-S** | 32.04 | 53.14 | K-S** |
| Percent of Population that is Hispanic | | | | | | |
| Mean | 13.93 | 25.27 | M-W** | 13.55 | 22.78 | M-W** |
| Variance | 93.68 | 161.27 | K-S** | 76.21 | 105.47 | K-S** |
| Percent of Population that is Black | | | | | | |
| Mean | 2.80 | 9.65 | M-W** | 2.81 | 8.35 | M-W** |
| Variance | 29.23 | 84.15 | K-S** | 23.33 | 55.65 | K-S** |
| Mean Gross Rent (in dollars per month) | | | | | | |
| Mean | 121.00 | 98.43 | M-W** | 123.41 | 103.14 | M-W** |
| Variance | 59.30 | 55.01 | K-S** | 48.84 | 44.82 | K-S** |
| Mean Contract Rent (in dollars per month) | | | | | | |
| Mean | 107.69 | 83.65 | M-W** | 109.44 | 88.98 | M-W** |
| Variance | 61.53 | 63.76 | K-S** | 48.84 | 49.91 | K-S** |

Table 4.7: (continued)

| Characteristic | Neighborhoods Searched In | | | Neighborhood Moved To | | |
|---|---|---|---|---|---|---|
| | Nonminority | Minority | Statistics | Nonminority | Minority | Statistics |
| Percent of Occupied Units that are Owner Occupied | | | | | | |
| Mean | 18.13 | 14.17 | M-W** | 18.75 | 14.75 | M-W** |
| Variance | 22.81 | 15.44 | K-S** | 20.34 | 12.18 | K-S** |
| Percent of Occupied Rental Units with all Plumbing Facilities, Direct Access, Complete Kitchen Facilities, and Heating | | | | | | |
| Mean | 96.03 | 92.41 | M-W** | 96.25 | 93.36 | M-W** |
| Variance | 14.82 | 26.92 | K-S** | 10.89 | 19.36 | K-S** |
| Percent of Units Built Before 1939 | | | | | | |
| Mean | 13.09 | 22.09 | M-W** | 12.78 | 20.15 | M-W** |
| Variance | 143.75 | 232.48 | K-S* | 111.09 | 163.33 | K-S** |
| Sample Size | (316) | (197) | | (214) | (137) | |

*Sample:* The data on search neighborhoods includes households searching in five or fewer neighborhoods during either the second six-month period and/or the second twelve-month period of the experiment; the data on move neighborhoods includes households moving at any time during the experiment. Both samples include only those households paying full-market rents in non-subsidized rental housing.

*Note:* For both the Mann-Whitney U Test (M-W) and for the Kolmogrov-Smirnov Test (K-S), significance is based on a two-tailed test.

*Statistic significant at .05 level.

**Statistic significant at .01 level.

Table 4.8: Mean Neighborhood Characteristics of Origin, Search, and Move Neighborhoods for Nonminorities and Minorities and Tests of the Similarity Between the Distributions of Origin and Search Neighborhoods, Search and Move Neighborhoods, and Origin and Move Neighborhoods for Nonminorities and Minorities in Pittsburgh

| | Origin Neighborhood | Search Neighborhoods | Move Neighborhood | Origin vs. Search[a] | Search vs. Move[a] | Origin vs. Move[a] |
|---|---|---|---|---|---|---|
| Mean Household Income (in $1,000) | | | | | | |
| Nonminority | 9.89 | 10.42 | 10.17 | 3.57** | 2.81** | 2.38** |
| Minority | 8.67 | 9.03 | 8.87 | 1.96* | 1.69+ | 0.93 |
| Percent of Households Below Poverty Line | | | | | | |
| Nonminority | 10.23 | 8.93 | 9.17 | 3.78** | 0.91 | 3.38** |
| Minority | 17.96 | 16.29 | 16.49 | 2.05* | 0.57 | 1.55 |
| Percent of Population that is Hispanic | | | | | | |
| Nonminority | --- | --- | --- | --- | --- | --- |
| Minority | --- | --- | --- | --- | --- | --- |
| Percent of Population that is Black | | | | | | |
| Nonminority | 7.94 | 6.71 | 6.22 | 1.52 | 1.54 | 2.38* |
| Minority | 45.76 | 34.06 | 38.96 | 3.18** | 1.93* | 1.87+ |
| Percent of Occupied Units that are Owner Occupied | | | | | | |
| Nonminority | 13.05 | 13.95 | 13.77 | 4.37** | 1.96* | 3.11** |
| Minority | 11.82 | 12.73 | 12.60 | 2.57** | 2.24* | 1.59 |
| Percent of Units Built Before 1939 | | | | | | |
| Nonminority | 72.99 | 69.62 | 71.23 | 3.90** | 3.81** | 1.83 |
| Minority | 72.38 | 69.86 | 70.54 | 2.11* | 1.39 | 0.66 |

Table 4.8: (continued)

| | Origin Neighborhood | Search Neighborhoods | Move Neighborhood | Origin vs. Search[a] | Search vs. Move[a] | Origin vs. Move[a] |
|---|---|---|---|---|---|---|
| Percent of Occupied Rental Units with all Plumbing Facilities, Direct Access, Complete Kitchen Facilities, and Heating | | | | | | |
| Nonminority | 91.02 | 91.09 | 91.25 | 0.19 | 0.82 | 0.49 |
| Minority | 88.44 | 87.94 | 89.80 | 0.63 | 1.00 | 1.97* |
| Mean Gross Rent (in dollars per month) | | | | | | |
| Nonminority | 94.75 | 97.51 | 96.55 | 2.82** | 2.70** | 1.88+ |
| Minority | 94.15 | 98.10 | 96.07 | 2.44* | 2.28* | 0.68 |
| Mean Contract Rent (in dollars per month) | | | | | | |
| Nonminority | 77.97 | 80.50 | 79.21 | 2.47** | 3.03** | 1.19 |
| Minority | 75.73 | 81.90 | 79.52 | 2.94** | 2.42* | 1.36 |
| Sample Size | | | | | | |
| Nonminority | (276) | (203) | (279) | (197) | (200) | (274) |
| Minority | (82) | (63) | (84) | (63) | (63) | (82) |

*Sample:* Households moving at any time during the experiment and paying full-market rents in nonsubsidized rental housing.

[a]The test of the null hypothesis of no difference is based on a two-tailed Wilcoxon matched-pairs signed-ranks Z statistic.

+Statistic significant at .10 level.
*Statistic significant at .05 level.
**Statistic significant at .01 level.

Table 4.9: Mean Neighborhood Characteristics of Origin, Search, and Move Neighborhoods for Nonminorities and Minorities and Tests of the Similarity Between the Distributions of Origin and Search Neighborhoods, Search and Move Neighborhoods, and Origin and Move Neighborhoods for Nonminorities and Minorities in Phoenix

| | Origin Neighborhood | Search Neighborhoods | Move Neighborhood | Origin vs. Search[a] | Search vs. Move[a] | Origin vs. Move[a] |
|---|---|---|---|---|---|---|
| Mean Household Income (in $1,000) | | | | | | |
| Nonminority | 10.50 | 11.05 | 10.92 | 4.06** | 1.53 | 2.84** |
| Minority | 8.65 | 9.35 | 8.99 | 5.38** | 3.10** | 2.98** |
| Percent of Households Below Poverty Line | | | | | | |
| Nonminority | 10.76 | 9.72 | 9.65 | 2.65** | 0.23 | 3.03** |
| Minority | 18.94 | 15.17 | 16.65 | 5.91** | 2.61** | 4.02** |
| Percent of Population that is Hispanic | | | | | | |
| Nonminority | 15.04 | 13.95 | 13.93 | 2.11* | 0.24 | 1.98* |
| Minority | 27.81 | 23.23 | 25.27 | 5.22** | 2.40* | 3.26** |
| Percent of Population that is Black | | | | | | |
| Nonminority | 3.73 | 3.16 | 2.80 | 2.26* | 1.17 | 3.43** |
| Minority | 10.75 | 8.58 | 9.65 | 4.61** | 2.63** | 2.15* |
| Percent of Occupied Units that are Owner Occupied | | | | | | |
| Nonminority | 17.27 | 18.52 | 18.13 | 4.72** | 1.90+ | 3.29** |
| Minority | 13.61 | 14.76 | 14.17 | 5.00** | 3.02** | 2.95** |
| Percent of Units Built Before 1939 | | | | | | |
| Nonminority | 16.10 | 12.93 | 13.09 | 3.77** | 0.19 | 3.56** |
| Minority | 25.27 | 19.15 | 22.09 | 5.30** | 2.72** | 2.97** |

Table 4.9: (continued)

| | Origin Neighborhood | Search Neighborhoods | Move Neighborhood | Origin vs. Search[a] | Search vs. Move[a] | Origin vs. Move[a] |
|---|---|---|---|---|---|---|
| Percent of Occupied Rental Units with all Plumbing Facilities, Direct Access, Complete Kitchen Facilities, and Heating | | | | | | |
| Nonminority | 95.61 | 95.95 | 96.03 | 0.50 | 0.62 | 1.03 |
| Minority | 91.43 | 93.35 | 92.41 | 5.61** | 2.90** | 2.79** |
| Mean Gross Rent (in dollars per month) | | | | | | |
| Nonminority | 116.61 | 122.86 | 120.95 | 3.93** | 1.58 | 3.10** |
| Minority | 94.11 | 103.31 | 98.43 | 5.66** | 3.13** | 3.41** |
| Mean Contract Rent (in dollars per month) | | | | | | |
| Nonminority | 104.52 | 108.34 | 107.69 | 2.67** | 0.41 | 2.40* |
| Minority | 79.44 | 88.12 | 83.65 | 5.21** | 3.13** | 2.58** |
| Sample Size | | | | | | |
| Nonminority | (310) | (241) | (316) | (235) | (239) | (310) |
| Minority | (197) | (165) | (197) | (163) | (164) | (196) |

*Sample:* Households moving at any time during the experiment and paying full-market rents in nonsubsidized rental housing.

[a]The test of the null hypothesis of no difference is based on a two-tailed Wilcoxon matched-pairs signed-ranks Z statistic.

+Statistic significant at .10 level.
*Statistic significant at .05 level.
**Statistic significant at .01 level.

Racial differences in search

Examining the results for Pittsburgh first (Table 4.8), we see that the characteristics of neighborhoods searched in for both groups generally differ from the characteristics of the origin neighborhood. Likewise, the characteristics of the neighborhoods searched in generally differ from the characteristics of the neighborhoods moved to for both groups of households -- the characteristics of the neighborhoods moved to tending toward those of the neighborhoods of origin. Indeed, in six out of eight comparisons, the characteristics of the neighborhoods moved to for minority households are not statistically different from those of the neighborhood of origin (in the case of nonminority, three characteristics show no statistical difference).

In Phoenix (Table 4.9), as in Pittsburgh, the data indicate a significant difference for both groups of households between the characteristics of the neighborhood of origin and the characteristics of the neighborhoods searched in. Unlike the results for Pittsburgh, the results for nonminority households in Phoenix indicate no statistically significant difference between the neighborhoods searched in and the neighborhoods moved to; for minority households, however, Table 4.9 indicates a consistent difference between neighborhoods searched in and moved to. Finally, we see that both groups of households move to neighborhoods with characteristics that are statistically different from the characteristics of their neighborhood of origin, although the characteristics of the neighborhood moved to are not as dissimilar from the characteristics of the neighborhood of origin as are the neighborhoods searched in.

## SUMMARY AND CONCLUSIONS

This paper briefly reviewed the implications of optimal housing search models for the behavior of minority households. The theoretical discussion showed that the models imply a lower optimal level of search and more restricted search areas for minority households. The third section tested these implications. Minority households were found to behave in ways consistent with the implications of higher search costs (i.e., positive values of $\delta$) facing these households. In addition, minority households were found to search and move to neighborhoods with characteristics different from those of nonminority households. Finally, a comparison of the characteristics of origin, search and move neighborhoods indicated a significant difference between the two groups in the sequential change in neighborhood characteristics.

The results contained in the third section indicate that minorities may be selecting dwelling units that are of a lower level of optimality than otherwise comparable households. This might explain the generally higher probability of mobility among minority households. If so, more complex models which explicitly account for differences in the rate of time preference, degree of risk aversion and expected length of tenure need to be estimated. If the racial differences in search and mobility presented in this chapter are rational reactions by minority households to the existence of prejudice, the final implication of the search models (i.e., the failure of self-correcting market mechanisms) for the issue of equity in housing markets is indeed troubling.

## NOTES

1. Comments and extensions to this work are contained in Shapiro (1974), Lee (1976), and Williams (1977).

2. Vidal (1978) and McCarthy (1980) do report some comparisons between minority and nonminority households.

3. Goodman (1974, 1975) outlines such models.

4. Preliminary results of this research were presented in a paper at the Conference on the Housing Choices of Low Income Families, Washington, D.C., March 9, 1979.

5. Obviously, representing the flow of housing services by a one-dimensional measure is a simplification which may understate the household's true gain from relocation.

6. For a statement of these axioms, see Phlips (1974).

7. This section is based on Courant (1978).

8. If search without recall is assumed, the optimal amount of search will be reduced.

9. For further information on the Housing Allowance Demand Experiment, see Struyk and Bendick, eds. (1981).

10. Iterative techniques can be employed for utility functions without analytic solutions, but such techniques can be costly.

11. Given the controversy which still surrounds the values of the demand elasticities, we would prefer not to restrict such parameters a priori, but rather, to select a function which allows these parameters to be estimated.

12. In order to assess the empirical efficacy of the Stone-Geary's characteristics and implications, a second functional form of the indifference map has been estimated for comparative purposes. This second function, the Cobb-Douglas, is a more restrictive form than the Stone-Geary, but it requires less information to estimate. It would be useful to know if results from alternative representations of household preferences in behavioral models differ substantially.

13. In Pittsburgh, black households are identified as minority; in Phoenix, black and hispanic households are identified as minority.

14. Similar results are reported in Lake (1980).

15. Data from the 1970 Census as well as discussions with local housing experts were employed to divide Allegheny County (Pittsburgh) and Maricopa County (Phoenix) into 196 and 20 neighborhoods, respectively. The divisions were based on local perceptions and an attempt to maximize within neighborhood homogeneity and between neighborhood heterogeneity. In Pittsburgh, neighborhoods were usually comprised of between 2 and 5 census

tracts; in Phoenix, neighborhoods were usually comprised of slightly more census tracts. Search radius is computed from the centroid of the household's residence census tract (origin) to the centroid of the neighborhood searched by the household (destination). Data on the neighborhoods were not collected for the first six-month period of the experiment.

16. For the results, see Cronin (1982).

17. The estimated relationships between the measures of search and the gain from relocation are generally significant for both forms of the household's indifference map.

18. Recall, that alterations in neighborhood search behavior due to the expectation of discrimination have been held constant. This variable is often significant with the same sign as that given for the race of the household head. Therefore, the total differences between minority and nonminority search behavior are larger than indicated in Table 5.

19. Data on all neighborhoods searched were collected only for households searching in five or fewer neighborhoods (which comprise the vast majority of searchers). Households searching in more than five neighborhoods gave the total number of neighborhoods searched and location of the first, last and most often searched neighborhoods. Analysis of this data does not indicate any significant difference between these two groups of searchers among the measures of search or the characteristics of neighborhoods searched in.

## REFERENCES

Courant, P. (1978) "Racial prejudice in a model of the urban housing market", Journal of Urban Economics, 5, 329-345.

Cronin, F. J. (1980 ) "An economic model of intraurban search and residential relocation", Washington, D.C.: The Urban Institute, Working Paper 1510-3.

----- (1982) "The efficiency of housing search", Southern Economic Journal (forthcoming).

Goodman, John L. Jr. (1975) "Housing market information and optimal housing search", Washington, D.C.: The Urban Institute. Memorandum.

----- (1974) "The housing search process: a theoretical framework", Washington, D.C.: The Urban Institute, memorandum.

Hicks, J. R. (1956) A Revision of Demand Theory. Oxford: Clarendon Press.

Lake, R. W. (1979) "Housing search experiences of Black and White suburban home buyers." Prepared for seminar presentation at the Office of Policy Development and Research, U.S. Department of Housing and Urban Development, Washington, D.C.

Lee, C. H. and E. H. Warren (1976) "Rationing by seller's preference and racial price discrimination", Economic Inquiry, 14, 36-44.

Masson, R. (1973) "Costs of search and racial price discrimination", Western Economic Journal, 12, 167-86.
McCarthy, K. (1980) Housing Search and Consumption Adjustment. Santa Monica, California: The Rand Corporation, P-6473.
Phlips, L. (1974) Applied Consumption Analysis. Amsterdam: North-Holland.
Shapiro, D. (1974) "Costs of search and racial price discrimination." Economic Inquiry, 12, 423-27.
Struyk, R.J. and M.Bendick (eds.) (1981) Housing Vouchers for the Poor: Lessons from a National Experiment. Washington D.C.: The Urban Institute Press.
Vidal, A. (1978) "Draft report on the search behavior of Black households in Pittsburgh in the housing allowance demand experiment". Cambridge, Massachusetts: Abt Associates Inc.
Williams, W. (1977) "Racial price discrimination: a note", Economic Inquiry, 15, 147-50.

# 5

# SPATIAL ASPECTS OF RESIDENTIAL SEARCH

**James O. Huff**

As a household searches for a new residence, a sequence of vacant dwellings is visited or seen. Each dwelling has a location and the set of locations corresponding to the set of vacancies seen constitutes a search pattern which may be represented as a point map. Each of the models discussed in this paper is designed to generate a search pattern. The generating rules are the assumptions concerning the nature of the search process occurring within a spatially dispersed set of vacancies. The resulting patterns are summarized and compared in terms of measurements taken on salient or important characteristics of the point map.

The impetus for this study is an emerging concensus among those working on residential mobility that the residential search process leading to and eventually culminating in a relocation decision is a critical element in the larger residential mobility process. The residential search problem contains within it at least three separate but related topics or questions:

(1) The length of search. How many alternatives does a household consider before selecting a new residence and what decision rules are employed as a basis for accepting or rejecting alternatives?
(2) The role of information in the search process. How does a household gain and use information from different sources during the search process?
(3) The spatial aspects of search. Where will the household look for a new residence?

## THE RESEARCH CONTEXT

The existing literature on residential search with respect to the length of search, the role of information, and the spatial constraints on search is reviewed as a background to some specific models of spatial search.

*AUTHOR'S NOTE: Support from N.S.F. Grant Soc 77-27362 is gratefully acknowledged.*

 DOI: 10.4324/9781003182085-6

## The Length of Search

Ever since Rossi's (1955) original observations on the length of housing search, there has been a continuing fascination with the observation that households apparently consider a very small number of alternatives prior to the selection of a new residence. A third of all homebuyers and almost half of all renters consider only one possibility (Barret, 1973; Hempel, 1969). The observation is bothersome because the selection of a new residence is obviously an important decision which is made within the context of a large and often very diverse set of unknown housing possibilities. Under such conditions, we would expect that the selection of a new residence would occur only after a much more extensive survey of the alternatives than is apparently the case.

Several possible explanations for the observed search behavior have been suggested. The first possibility is that households are satisficers in the sense that they are searching for something better than they have and will stop searching when an alternative meeting certain minimal conditions is found (Wolpert, 1965). One explanation for satisficing behavior in the search process is that people may be quite risk averse when it comes to the selection of a new residence. Particularly when the housing market is tight, risk averse households will engage in limited search culminating in a housing choice which is satisfactory but often far from optimal (Smith et al., 1979). The second alternative is that the actual or perceived cost of search relative to available resources is very high. The implication is that search costs act as important constraints on the length of search. The third possibility is that the questions commonly asked by those doing empirical research on this problem provide a somewhat misleading image of the search process in that they evoke only the set of alternatives which are seriously considered by the household. The consumer behavior literature indicates that the evoked set of alternatives is indeed critical in the choice process but it may well comprise a very small subset of the awareness set (Howard, 1977). It is quite possible that the household becomes aware of many more vacancies prior to and during the active stages of search than is evidenced by actual visits to selected alternatives in the evoked set.

The first two explanations pertaining to satisficing and search costs can be accomodated within a Stigler-type search model (Stigler, 1961; 1962) which is designed to predict both the length of search and the conditions under which the household will stop searching. Although the Stigler model was originally designed to characterize job search behavior, it has been successfully adapted to the residential mobility and migration situations by Flowerdew (1973) and Smith et al. (1979) in the case of residential search, and by Goodman (1980) in the case of destination search resulting in migration.

Goodman, in particular, has identified a series of testable propositions concerning the optimal length of search and the nature of the stopping rule employed by households given that the search model accurately describes search behavior. Three of the eight propositions may be directly related to the general explanations of limited search given above. If G is the distribution of net gains to the household for all vacancies (or areas) in the domain of search (gains relative to the household's current living condition) then the following propositions (as adopted from Goodman) should be true:

(1) Given the known costs of search and the expected utility or net gain from search, there is some minimum net gain, $\bar{g}$, such that any vacancy found with a net gain g greater than or equal to $\bar{g}$ will be selected even though further search might uncover a better alternative;

(2) the greater the variance of the G distribution, the greater is the expected number of units searched; and

(3) the lower the costs of search relative to the expected value of G, the greater is the expected number of units searched.

Propositions 2 and 3 stated above appear to be supported by empirical studies of search behavior. If we assume that the distribution of net gains for the average renter, $G_R$, has a smaller variance than the distribution of net gains for the average homebuyer, $G_O$, then proposition 2 implies that renters, on the average, investigate fewer alternatives before making a choice than is the case for homebuyers. Empirical studies of search behavior have consistently found significant differences in the length of search exhibited by homebuyers and renters (Michelson, 1977; Rossi, 1955; Speare et al., 1975). Propositions 2 and 3 are also supported by Meyer's recent study of choice set formation under controlled laboratory conditions (Meyer, 1980). His study indicates that "the size of an individual's choice set is likely to be largely a function of the amount of variance which exists in utilities within the population of alternatives and the amount of time and money available for searching" (Meyer, 1980:31).

Much of the direct support for proposition 3 has arisen out of the empirical work done on the search behavior of low income and minority populations. The studies of low income populations indicate, not unexpectedly, that their search behavior is often severely constrained by resource limitations in the form of poor access to private transportation and minimal amounts of previously uncommitted time. The relative cost of search, therefore, tends to be limited both in the number of alternatives considered and in the spatial extent of search (Cronin, 1979; McCarthy, 1979; Weinberg et al. 1977). Because of the barriers to search accompanying overt or anticipated discrimination, it can also be argued that minority populations also face higher expected search costs than their non-minority counterparts--particularly in those areas which are either explicitly or implicitly proscribed on the basis of the household's race. Under the above conditions, we would expect that minority households investigate fewer alternatives and that these alternatives are concentrated in minority areas as is the case (Cronin, 1979; McCarthy, 1979).

Although the cost of search is incorporated in the empirical and theoretical discussions of the length of search and the decision to stop searching, very little work to date has been done on the relative costs of using different information sources. Only McCarthy (1979), Clark and Smith (1979) and Smith and Clark (1980) appear to have investigated the relationship between the costs of information acquired from different sources and the length of search. This line of enquiry provides an essential link between the length of search and information sources discussed in the following section. It may also help to shed some light on the processes leading to the formation of the evoked set which is as yet an open question.

## The Role of Different Information Sources In The Search Process

Although questions of information channel utilization and effectiveness have been of continued concern to those working on residential search, the existing studies provide only fragmentary insights into how households acquire and use search related information. The descriptive studies have been largely concerned with the relative importance of different information sources as measured in terms of the number of households employing each source during different phases of the search process, but even at this very general level, we still do not have adequate explanations for the large discrepancies in the findings when the various studies are compared (see Clark and Smith, 1979, for a review).

An important question which is as yet only partially answered concerns the relationship between household characteristics such as socio-economic status, race, or tenure and the types of information sources employed. Barret (1973) in his Toronto study provides indirect evidence in support of Rossi's (1955) observation that lower status households tend to rely more heavily on personalized information sources such as friends and relatives. This observation is also supported by preliminary findings from the housing allowance studies which also indicate that minority groups rely less heavily upon newspapers and driving around as sources of information (Cronin, 1979; McCarthy, 1979; Weinberg et al., 1977).

Only Hempel (1969) and the later theoretical work of Clark and Smith (1979) focus on issues surrounding the sequence of information sources used during search. These studies indicate that the early stages of search are often characterized by a greater dependency on newspaper advertisements and "driving around" while later stages are characterized by an increasing dependence on real estate agents in the case of homebuyers.

The investigation of the information content of messages supplied by different sources has also been neglected. Notable exceptions to this observation are Burke, et al.'s (1979) study concerning buyer cognition of information source content, and Smith and Clark's (1979) simulation model which explicitly treats the effects of information content upon the sequence of sources used, the length of search, and the areas searched by the household.

Within this small set of studies on information sources, an even smaller number have investigated the spatial aspects of the problem. Barrett (1973) and Michelson (1977) find that homebuyers moving to different areas of Toronto tend to differ in their use of sources at least in the initial phases of search with "driving around" predominating in the peripheral areas, real estate agents dominating in the high turnover areas of the city and to the limited extent that personal contacts were employed, they were concentrated in the low income areas, thus supporting Rossi's earlier observations. Michelson also finds that newspapers play a greater role in the search process when the move is from one general area of the city to another (i.e., downtown to suburbs).

Palm restricts her attention to a single information source--real estate agents--and focuses upon the existence of spatial bias in the information provided by agents. Although she does not have data on the content of realtor messages, she does find that realtor recommendations for hypothetical or exemplar households have marked spatial biases indicating that agents tend to overrepresent the possibilities of success in those areas surrounding their own office (Palm, 1976).

The theoretical implications of real or perceived discrimination practices by brokerage agents and/or sellers have been discussed by Courant (1978) and Yinger (1978). They show that it may be in the broker's best interest to steer minority households away from predominantly white areas of the city, thus leading to minority search strategies which avoid areas of potential discrimination and resulting in housing choices in minority areas even in instances where prices (for comparable housing) may be significantly higher in the minority area.

These studies have made important contributions to our understanding of the spatial biases in the household's information sources but they of necessity leave many important questions unanswered. It would be useful to know, for example, how the relative supply of housing in any one area of the city tends to affect the set of information sources employed by the seller or apartment owner. Although much more attention must be devoted to the spatial filters imposed by real estate agents, this work would be nicely complemented by studies of the spatial biases inherent in other sources such as newspapers or driving around. In this regard, Schneider (1975) has suggested an interesting spatial search model which is particularly appropriate for search involving information collected by driving around or walking around an area.

## The Spatial Context of Search

The small number of studies which focus on the spatial aspects of household search behavior may be divided into two groups: those that describe search in terms of the areas searched (Clark and Smith,

1979; Smith et al., 1979) and those describing characteristics of the point pattern consisting of the locations associated with the vacancies visited by the household during search (Barrett, 1973; Brown and Holmes, 1971). The two approaches are largely complementary in that the area based approach is concerned with occupancy or membership measures and the point pattern approach tends to employ distance and/or directional measures but both are ultimately interested in where a household searches for a new residence.

Although the spatial aspects of search have received less attention than the other two search related issues already discussed, we do have some evidence to suggest that the search pattern is influenced by the socio-economic status and the race of the individual with low status and minority populations exhibiting comparatively localized search patterns (Barrett, 1973; Brown and Holmes, 1971; McCarthy, 1979). Another closely related result is that inner-city residents relative to those living on the periphery exhibit more localized or spatially concentrated search behavior. Two competing explanations exist for the observed differences in the search behavior of the different subpopulations in question. The first argument is that the search pattern reflects the distribution of vacancies in the opportunity set. Low priced housing and/or housing in predominantly minority areas is spatially concentrated within the city. Therefore, it should not be surprising that low status households and minorities facing potential discrimination outside of minority areas exhibit spatially concentrated search behavior. The second argument is based upon the observation that households tend to search in already familiar areas (Barrett, 1973). Inner-city/low status/minority households are likely to have more limited awareness spaces in part because of the restricted activity spaces of these individuals (Wolpert, 1965). We would therefore expect that these subpopulations would exhibit spatially constrained search behavior. Again, these two arguments are largely complementary with biases in the awareness space of the individual reinforcing in many instances, the already existing spatial biases in the opportunity set faced by the household.

The search model outlined in this paper draws upon the opportunity set argument and the awareness space concept in conjunction with several ideas pertaining to the relative cost of gaining and using information. The search procedure described by the model is a two-step process consisting of first, the selection of an area where search is to be concentrated, and second, the actual selection of vacancies which are within the targeted area and are known to be members of the possibility set N. The spatial disequilibrium residential search model falls within this general class of models (Smith et al., 1979). The decision to search in an area is assumed to be a function of the household's expected utility of searching in this area. The model is attractive because the choice of an area and the decision to stop searching are shown to be functions of variables which almost certainly have a bearing on the search process -- variables such as the household's preferences and budget constraints, the household's beliefs concerning the household's degree of risk aversion.

In theory, the spatial disequilibrium model is a learning model since the household bases its expectations upon beliefs about the market; and as search proceeds, these beliefs are modified as new information is acquired. Unfortunately, the nature of the updating procedure is as yet unspecified, although Clark and Smith (1979) have begun to investigate the effects of different information sources and their associated costs upon the updating of beliefs using a simple Bayesian model. Until the updating problem is solved, the model will remain an interesting theoretical exercise. In its present form, it cannot be used to predict the household's search pattern unless we have a continuous update on the household's beliefs concerning the distribution of housing utilities in each area.

## A MODEL OF SPATIAL SEARCH

An approach which draws upon some of the ideas in the spatial disequilibrium model but which is not dependent upon a knowledge of household beliefs is therefore a more normative model in that households looking for the same kinds of residences (i.e., households with similar possibility sets) tend to generate the same kinds of search patterns. The model outlined in this paper assumes that the household uses one or more sources to acquire indirect information on selected attributes of vacancies. Certain vacancies are rejected solely on the basis of the indirect information provided while others are considered to be viable possibilities. The household proceeds to visit vacancies in this possibility set until a new residence is found or the household stops searching.

Members of the possibility set which are located in areas of the city with relatively low expected search costs are assumed to have a higher probability of being identified and visited than do members located in higher search cost areas. The search patterns generated by the model are therefore not only a function of the underlying distribution of vacancies in the possibility set but also a function of spatial biases in the actual search process. Since the expected cost of search is likely to be as difficult to measure directly as the expected utility of search in the spatial disequilibrium model, an additional set of assumptions is introduced which relate the expected cost of search to area related variables which can be more easily measured and whose values can be determined prior to the beginning of the search process.

The inferences derived from the model are in terms of measurements taken on the search patterns generated by the model. Two general classes of measurements are used to summarize the patterns:

(1) occupancy or membership measurements -- vacancies which are observed are assigned to one of n mutually exclusive areas or neighborhoods in the city,
(2) distance and/or directional measurements
 i) taken over the set of observed vacancies
 ii) taken with respect to an external reference point (e.g. present location, workplace, realtor location).

Both classes of measurements may be further differentiated depending upon the decision to include the ordering of observed vacancies as an aspect of the measurement problem. For example, if the ordering of the sequence is deemed unimportant, then a search pattern might be summarized by the number of vacancies seen in each small area of the city. If ordering is maintained, then the summary measure would be the ordered sequence of small areas corresponding to the areas in which the observed vacancies are located. If measurements are in terms of distance, a distribution of interpoint distances for the set of observed vacancies could be generated; whereas, a distance measure over the ordered set might be the ordered sequence of distances between the first vacancy observed and each observed vacancy thereafter.

Residential search is assumed to occur under the following general conditions:

(1) Each household has a conditional utility $V\{p, Y-M, X\}$ for housing with price, p, of non-housing commodities, income, Y, price or rent, M of a given dwelling, and a set of attributes, X, of the house.
(2) A reservation utility level $V^c$ exists for a given household such that a vacancy i having a utility, $V^i \geq V^c$ would be chosen if seen. To simplify the problem, it is assumed that $V^c$ remains constant during the search process. Let A be the set of all vacancies i with $V^i \geq V^c$.
(3) An information source S, exists which provides information on selected attributes of a given vacancy i such that a household using this source to gain information about a vacancy i is aware of some but not all of the elements in the set $\{M_i, x_{i1},...x_{in}\}$ which completely describes the attributes of the vacancy i where $B_i = \{L_i \mid S\}$ is the subset of attributes defining i for which S provides information. Note that S may or may not accurately

report information on a given attribute of i. The "value" of a given information source is therefore a function not only of the "kinds" of information provided but also is a function of the "accuracy" of the information. Furthermore, the information source may be biased in the sense that all vacancies do not have an equal chance of being reported by S and the accuracy may also vary over the set of reported vacancies.

(4) Given a sequence of vacancies for which S provides information, the household has a decision rule which is used to assign each vacancy in H to one of two disjoint sets: the set of vacancies, R, which can be rejected solely on the basis of information provided on B; and the set of vacancies, N, which cannot be rejected without direct investigation on the part of the household.

(5) A vacancy which is a member of N is assumed to have a constant probability K of being chosen by the household if it is visited by a household. The search process may also be terminated because the household decides to stop searching and to remain in its present location. The probability of terminating an unsuccessful search after $\omega j$, the $j^{\underline{th}}$ vacancy seen, is assumed to be a constant e. Given that a household has already seen j-1 vacancies, the probability that it decides to continue searching is (1-K-e).

(6) Given that the household is still searching, the probability that $\omega j$ is a specific vacancy, $x_i$, contained in $N_i$, the set of vacancies in area i which are members of N, is assumed to be inversely proportional to the expected cost, C(i, j), of finding a new residence in area i. The conditional probability, p(i, j), that $\omega j$ is located in area i is therefore:

$$p\,(i, j) = n_i \ C(i, j)^{-1} / \sum_k n_k \ C(k, j)^{-1} \qquad (1)$$

where $n_k$ is the number of vacancies in $N_k$.

These six conditions define the basic structure of an area based model. The model in its general form is not operational, however, because the decision rule in condition 4 is as yet unspecified. Two different types of decision rules are explored. The first is based upon minimum requirements criteria, and the second uses a direct assessment of the probability that a vacancy in H will have a utility which greater than the reservation utility $V^c$.

Decision Rules For Area Based Search Models

The minimum requirements decision rule presumes that the household will not accept a house which has less than $\hat{x}_i$ of a given attribute i, $\forall$ i, nor will it pay more than $\hat{m}$ for any house no matter how attractive that house may be on other attributes. Since the information source S only provides information on the values for a limited set of attributes, $B = \{b_j\}$, a vacancy i will be rejected if for any known attribute, $b_j < \hat{x}_j$.

Thus we have a situation similar to that represented Figure 5.1. In order to determine the magnitude of each minimum requirement, $\hat{x}$ we could:

(1) ask the household (much as a realtor does) what the household's minimum requirements or constraints are; or
(2) infer them from an experiment in which $x_i$ would be that

$$\text{value of } x_i \text{ for which } V\left\{x_{j \neq i}^{\max},\ x_i\right\} = V^c.$$

The advantage of the minimum requirements decision rule, aside from its conceptual simplicity, is that it can be motivated by a psychological theory of decision-making. Tversky (1972) asserts that

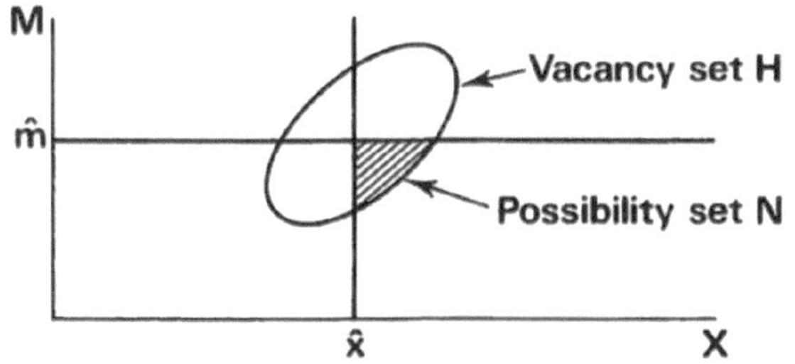


Figure 5.1. The possibility set.

individuals, when faced with a complex choice, use a sequential elimination process (the elimination by aspects model) in an effort to sort out those alternatives which do not have certain qualities or attributes where the attributes are ordered in terms of their relative importance to the individual. Although the minimum requirements decision rule does not necessarily invoke the strong ordering assumption contained in the elimination by aspects model, the final outcome (the set N of vacancies which cannot be rejected without further investigation) is the same under the two procedures.

An alternative to the minimum requirements decision rule is a decision procedure based upon the household's subjective assessment of the chances of success given incomplete information on a particular vacancy. A vacancy is included in the choice set N if when provided with information on a subset of attributes, the household's subjective assessment of the probability of success is greater than some value $\hat{K}$, $0 \leq \hat{K} \leq 1$. In other words, a household might decide that a vacancy $i$ is worth going to see if it felt that there was a better than 10% chance that the vacancy would be chosen once seen where $\hat{K}$ is 10 percent in this instance. The associated choice set, N, would therefore be all vacancies which have a subjective probability of success greater than 10 percent given information on $B_i$.

If a household selects a cut-off level $\hat{K}$ then there is an associated expected probability of success, K, where K is necessarily greater than $\hat{K}$ since all members of N have a subjective probability of success greater than $\hat{K}$. In theory, the household selects the cut-off value, $\hat{K}$, and the associated expected probability of success K, on the basis of the cost of acquiring indirect information on vacancies from an information source S relative to the cost of actually visiting vacancies. As the cost of visiting vacancies increases relative to the cost of acquiring indirect information, the cut-off value will tend to increase reflecting the household's willingness to increase the number of vacancies investigated indirectly through S so as to reduce the expected number of vacancies which the household will visit before a choice is made.

No matter which decision rule is used, the fact that most households use indirect information to select a potentially small subset N of likely possibilities will almost certainly prove to be a critical element in any of our spatial search models. If the spatial distribution of vacancies in the choice set N does not correspond to the spatial distribution of a similar number of vacancies randomly sampled from the set of all existing vacancies then the decision to investigate only those vacancies which are members of N will necessarily result in a spatially biased search pattern.

Figure 5.2 provides two examples of choice sets faced by prospective home buyers in the western portion of the San Fernando Valley in Los Angeles during the first two weeks of April, 1979. In the first instance (A) the assumption is that the household in question is searching within the set of all vacancies which have a listed price of less than \$80,000 and have at least three bedrooms. The second

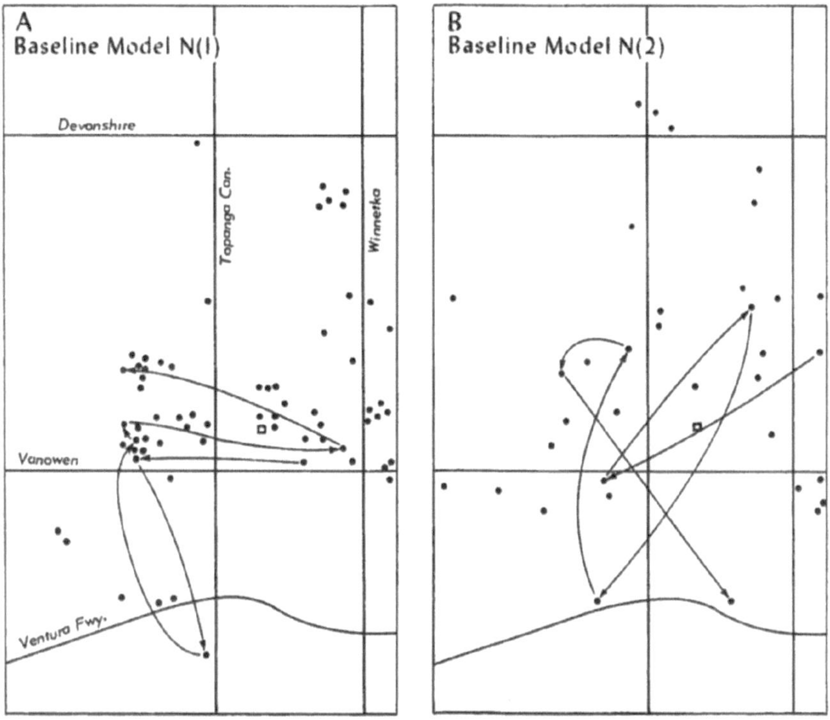


Figure 5.2. A simulated search sequence under baseline conditions. Each dot on the map represents a vacancy in the potential choice set. The potential set, N(1), shown in A, consists of all vacancies with listed prices less than $80,000 and three or more bedrooms. The potential choice set, N(2), shown in B, consists of all vacancies with listed prices less than $90,000, four or more bedrooms, and at least 1.75 baths. The reference point is indicated by □.

choice set, N(2), consists of all vacancies which have a listed price of less than $90,000, 4 or more bedrooms and at least 1 3/4 baths.

Even a visual comparison of the two choice sets indicates that the vacancies in N(1) are more clustered spatially than are the vacancies in N(2). All else being equal, we would therefore expect the observed search patterns generated by households searching within N(1) to be more clustered than for comparable patterns generated by households searching within N(2). The baseline case outlined in the next section makes explicit the spatial biases in the search pattern arising from differences in the underlying distribution of vacancies in the choice set.

## A SIMPLE OPERATIONAL AREA BASED SEARCH MODEL

For comparative purposes it is useful to describe the properties of the search pattern which arises when search costs do not vary from one area to another and are independent of past search behavior such that C(i, j) in equation (1) is some constant c. When the cost of search is constant, the probability that a given vacancy, x, is the $j^{\underline{th}}$ vacancy seen in the ordered set W of all vacancies seen by a household is:

$$P_r \{w_j = x\} = \begin{cases} 1/n, & x \in N \\ 0, & \text{otherwise} \end{cases}$$

where n is the number of vacancies in N. Simulated search patterns generated under the baseline condition given above are illustrated in Figure 5.2.

The six basic properties of the search pattern generated under the simple conditions are as follows:

(1) If $\omega$ vacancies are seen during a search sequence $\omega$ then the number of vacancies seen in a given area i, $z_i$, is binomially distributed with parameters $(p_i, \omega)$ where $p_i = n_i/n$, $n_i$ being the number of vacancies in area i which are members of the set N. The area probabilities for choice set N(1) are as shown in Figure 5.3(A). Each area is approximately 1.5 square miles.

(2) The length of search as measured in terms of the number of vacancies seen before terminating search is geometrically distributed with parameter $r = K+e$ implying that the expected length of search is $r^{-1}$.

(3) The expected number of areas visited during search is $\sum_i y_i$ where $y_i$ is the probability that at least one vacancy is visited in area i such that

$$y_i = 1 - (1 - p_i)^{r^{-1}}.$$

(4) If directional bias is measured in terms of the number of observed vacancies in sectors radiating out from a designated point, a, then the number of vacancies seen in a given sector i is binomially distributed with parameters $(p_i^*, \omega)$, $p_i^* = n_i^*/n$, $n_i^*$ being the number of vacancies in sector i which are members of N.

| A<br>Baseline Model N(I) | | | |
|---|---|---|---|
| $P_i$<br>M(F(j))<br>$S^2$ | | .015<br>.10<br>.11 | .076<br>.51<br>.68 |
| | | .015<br>.10<br>.11 | .030<br>.20<br>.23 |
| | .182<br>1.2<br>2.3 | .060<br>.40<br>.51 | .030<br>.20<br>.23 |
| | .197<br>1.3<br>2.5 | .091<br>.60<br>.80 | .212<br>1.4<br>2.8 |
| .030<br>.20<br>.23 | .045<br>.30<br>.38 | | |
| | | .015<br>.10<br>.11 | |

| B<br>Area Based Model N(I) | | | |
|---|---|---|---|
| $\pi$<br>M(F(j))<br>$S^2$ | | .006<br>.03<br>.03 | .059<br>.35<br>.33 |
| | | .008<br>.05<br>.04 | .024<br>.13<br>.12 |
| | .200<br>1.3<br>1.9 | .045<br>.26<br>.23 | .012<br>.07<br>.06 |
| | .224<br>1.5<br>2.1 | .109<br>.68<br>.69 | .272<br>2.0<br>2.7 |
| .013<br>.07<br>.07 | .029<br>.16<br>.15 | | |
| | | .004<br>.02<br>.02 | |

Figure 5.3. Area related summary statistics. The area probabilities represent the probability that $\omega_j$ occurs in the designated area under baseline conditions (Figure A) and that $\omega_j$ occurs in the designated area under conditions given in the area based model (Figure B). The expected number of vacancies seen in each area are also provided. The probability of success, K, in both models is set at .15, and the other parameter values for the area model are as follows:

$$\alpha = .6, \; c_D/c_s = 5, \; \beta_s = 1.0$$

(5) The expected distribution of distances between a point, a, and each of the vacancies in the ordered sequence $\omega$, of the dwellings seen by the household is summarized by the set of moments $\{m_i(a)\}_{i=1}^{\omega}$ such that

$$m_i(a) = 1/n \sum_j s^i(j)$$

where $s(j)$ is the distance between the $j^{th}$ vacancy in N and the reference point, a. The first three reference point moments for N(1) are given in Table 5.1.

(6) The $i^{th}$ moment of the distances between vacancies in $\omega$ is

$$m_i = 1/n\,(n-1) \sum_j \sum_k s^i\,(j, k),\ k \neq j$$

where $s(j, k)$ is the distance between the $j^{th}$ vacancy in N and the $k^{th}$ vacancy in N.

The first three interpoint moments for N(1) are 4.2, 22.9, and 146.1 under baseline conditions. The reference point, is as shown in Figure 5.2.

(7) In the ordered sequence, W, of vacancies seen, the location of $\omega_j$ is independent of the location of $\omega_i$ in W in the sense that

$$\Pr\{\omega_j = x \mid \omega_i = y\} = \Pr\{\omega_j = x\}, \text{ for all i and j.}$$

Although the characteristics of the search pattern outlined above are not exhaustive, they do summarize most of the important aspects of the pattern as identified by those attempting to describe and compare such patterns.

The above properties of the expected search pattern are based upon the assumption that sampling with replacement is occurring. A slight variation of the choice set model assuming sampling without replacement may be worth investigating particularly in those

Table 5.1: Reference Point Moments for N(1)[a]

| Moments | Baseline Model | Area Based Model[b] |
|---|---|---|
| $m^1(a)$ | 3.1 | 2.5 |
| $m^2(a)$ | 12.1 | 8.2 |
| $m_3(a)$ | 53.8 | 32.9 |

*Notes:* a. The reference point is designated by a "□" in Figures 5.2 and 5.4.

b. The parameters for the area based model are as given in Figure 5.3.

instances where N is small. If the household is sampling vacancies without replacement and $\omega$ vacancies are seen during a search sequence W then the number of vacancies seen in a given area i for example has a hyper-geometric distribution with paramenters $(n, n_i, \omega)$. In particular, if r of the first $K$ vacancies in W occurred in area i then the probability that the k + 1$^{st}$ member of W is also located in area i is

$$Pr\left\{\omega_{k+1} = x \epsilon \text{ area i}\right\} = (n_i - r)/(n - k);$$

whereas if sampling with replacement is occurring,

$$Pr\left\{\omega_{k+1} = x \epsilon \text{ area 1}\right\} = n_i/n.$$

The spatial implications of the simple model reflect the fact that the pattern of observed vacancies is solely a function of the distribution of vacancies in the possibility set -- vacancies which cannot be rejected given information on a limited set of attributes. Throughout the discussion it has been assumed that the household is randomly selecting possibilities from N; and by implication, spatial biases do not exist in the household's selection process. If the household does base its decision to investigate a vacancy upon the location of that vacancy, then the simple model provides a means of assessing the role of the household's selection process in shaping the resulting search pattern.

## AREA BASED SEARCH AS A MARKOV CHAIN

The general model as summarized in equation (1) and operationalized in the preceding section does not specify the nature of the cost function which is ostensibly a function of the area in which search occurs and the household's prior search experience. In this section, the expected search cost will be broken down into its component parts. Under certain simplifying conditions it is then shown that the search process may be described as a markov chain in which the probability of searching in a given area, j, is a function of the location of: the last vacancy seen; the indirect cost of identifying a member of the possibility set N which is located in the area j; and the direct cost of actually visiting that vacancy.

The expected cost of finding an acceptable vacancy in area i at any point j in the search sequence may be separated into an indirect cost of search, $C_I(i, j)$; a direct cost, $C_D(i, j)$; and a start-up cost, $C_E(i, j)$. The expected indirect cost is the cost associated with the generation of possibly acceptable vacancies in an area; the expected direct cost is the cost of actually visiting possibilities; and the expected start-up cost is the cost of shifting the focus of search to area j given that the previous vacancy seen was the result of search in another area.

If there is a fixed cost, $c_D$, of visiting a single vacancy then the expected direct cost of finding an acceptable vacancy is a constant, $C_D$; and

$$C_D = K^{-1} c_D$$

since $K^{-1}$ is the expected number of vacancies seen before a success.

The expected cost of initiating search in an area is assumed to be proportional to the expected cost of finding an acceptable vacancy in the area such that

$$C_E(i, j) = \alpha\, C(i, j).$$

If $C_s(i, j)$ is the expected cost of using an information source S to identify a member of $N_i$ and the household uses only one main source S to generate possibilities then the expected indirect cost of search is

$$C_I(i, j) = K^{-1}\, C_s(k, j).$$

If there is a constant cost, $c_s$, of using S to generate information on a single vacancy then:

$$C_s(i, j) = G(i, j)\, c_s$$

where G(i, j) is the expected number of vacancies generated by S before a member of $N_i$ is identified. If the household is searching for a new residence in area i then the probability that any given vacancy in the possibility set $N_i$ is identified by an information source S is assumed to be inversely proportional to the total number of vacancies in area i, $h_i$. The constant of proportionally, $\beta_s$, is a measure of the efficiency of S where $\beta$ is normally greater than or equal to one since a value of one would imply that S randomly selects vacancies from the set of all vacancies in area i. The expected number of vacancies generated by S before a member of $N_i$ is identified is therefore

$$G(i, j) = (\beta_s)^{-1}\, h_i/n_i$$

which implies that the expected indirect cost of searching for a vacancy in area i is

$$C_I(i) = (K\beta_s)^{-1} c_s\, h_i/n_i.$$

The total expected cost of searching for $\omega$ in area i may now be stated as a function of the density of possibilities in the area, $\rho_i = n_i/h_i$ and the area containing $\omega_{j-1}$:

$$C(i, j) = K^{-1} c_D + (K\beta_s\, \rho_i)^{-1} c_s + \xi(i, j)\alpha C(i, j) \qquad (2)$$

where: $\xi(i, j) = \begin{cases} 1 \text{ if } \omega_{j-1} \notin i \\ 0 \text{ if } \omega_{j-1} \in i \end{cases}$

Given that search is still continuing, it follows from (1) and (2) that the probability of searching for $\omega_j$ in area i is conditional upon the location of $\omega_{j-1}$ such that:

$$p(i, j) = \sum_k t_{ki} \, p(k, j-1) \tag{3}$$

where: $t_{ki} = \begin{cases} z_k / \sum_\ell (1-\alpha) z_\ell + \alpha z_k \text{ for } t_{kk} \\ (1-\alpha) z_i / \sum_\ell (1-\alpha) z_\ell + \alpha z_k \text{ otherwise} \end{cases}$

and: $z_i = n_i \, [(c_D \, \beta_s / c_s) + \rho_i^{-1}]^{-1}$

Two important implications follow immediately from the above result. The first is that the probability of searching in a given area varies only as a function of $\rho_i$ which in conjunction with $\beta_s$ is the relative concentration of viable possibilities in the stream of vacancies from an area for which S provides information. The only other parameters in the model which need to be estimated are the ratio of direct to indirect search costs, $c_D/c_S$, and $\alpha$ which is a measure of the importance of start-up costs relative to total expected search costs for an area. The second observation is that the search model describes a search process which is an absorbing markov chain with a state space consisting of 2M states where the first M states are absorbing states being the M areas from which the household chooses its new residence and the second M states are the transient states being the same M areas where the search for vacancies may be conducted. The associated transition matrix P has the following form:

$$P = \left( \begin{array}{c|c} I & 0 \\ \hline R & Q \end{array} \right)$$

Where I is an M by M identity matrix representing the absorbing states, R is an M by M matrix with $r_{ii}$ = K, the probability that search is concluded in area i given that a vacancy is seen in area i, and $r_{1i}$ = e assuming that area 1 is designated as the area containing the current residence. The matrix Q is also an MxM matrix with $q_{ki}$ being the probability that the next vacancy seen is in area i given that the previous visit was to vacancy in area k. It follows from (3) that

$$q_{ki} = (1-K-e) \, t_{ki}$$

Spatial aspects of search

The expected search pattern generated by the search cost model may be summarized by areas and distance related measures similar to those outlined for the baseline case. The fact that the model is an absorbing markov chain makes it possible to make explicit statements about the nature of the search patterns generated by the model. The area related results shall be stated without proof since they are given in any standard text on markov chains. Using Kemeny and Snell's 1960 notation, the area related results are as follows:

(1) The expected number of vacancies seen in area j prior to successfully completing the search process is $M(f(j) = \pi \ell_{\cdot j}$ where $\Pi = \{\pi_i\}$ is the vector of initial probabilities and $\ell_{\cdot j}$ is the $j^{\underline{th}}$ column vector of the matrix L such that $\pi_i = t_i/T_i$ and $L = (I-Q)^{-1}$. The vector of second moments is

$$M(f^2) = \pi L(2L_{dg} - I)$$

where $L_{dg}$ is a diagonal matrix with $\ell_{ii}$ as entries on the main diagonal.

(2) Given that the first vacancy seen by the household occurs in area i then the probability that the household finally concludes its search in area j is $b_{ij} = K\ell_{ij}$, $\ell_{ij}$ being an element of matrix L. The unconditional probability that the household ends its search in area j is

$$b_{\cdot j} = K\sum_i \pi_i \ell_{ij}.$$

(3) Given that a household begins searching in a new area i, $(1-q_{ii})^{-1}$ is the expected number of vacancies seen in that area before the household changes state and the associated variance of the distribution is $q_{ii}/(1-q_{ii})^2$.

(4) The expected number of areas searched by the household before completing its search given that search begins in area i is

$$M_i(t) = \hat{L}\xi$$

where is a MX1 column vector of 1s and

$$\hat{L} = (I-\hat{Q})^{-1}$$

with $$\hat{q}_{ij} = \begin{cases} q_{ij}/1-q_{ii}, & j \neq i \\ 0, & j = i. \end{cases}$$

The second moment of the distribution is

$$M_i(t^2) = (2\hat{L}_{dg}-I)\hat{L}\xi.$$

The initial probability of searching in area i as well as the mean and variance for the number of vacancies seen in area i are given in Figure 5.3(B). A simulated search pattern under area based conditions is also provided in Figure 5.4.

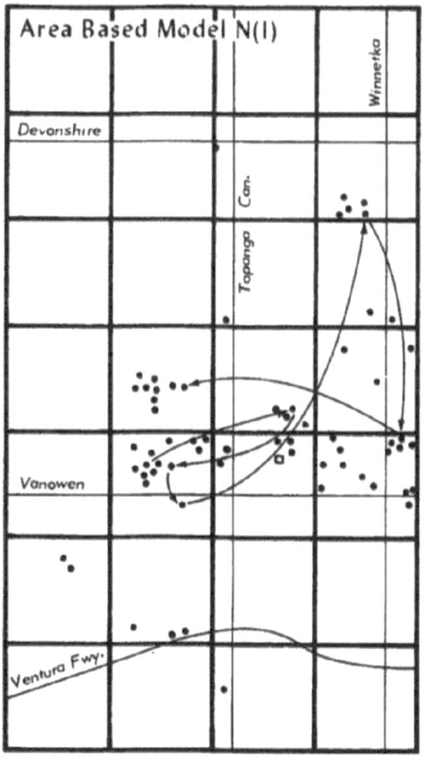


Figure 5.4. A simulated search sequence under conditions given in the area based model where:

$$K = .15,\ \alpha = .6,\ c_D/c_s = 5,\ \beta_s = 1.0$$

(5) Given that a reference point a in area j is the last vacancy seen in an uncompleted search sequence and a set of sectors radiating out from a are defined then the probability that the next vacancy seen will be sector i is

$$\theta_a(i) = \sum_k \Pi\ell_{\cdot k}\, n(k, i)/n_k.$$

The second moment of the distribution is

$$M(\theta_a^2(i)) = \sum_k M(f^2(k))\ (n(k, i)/n_k)^2$$

where M $(f^2(k))$ is as given in the first area related result.

(6) The $i^{\underline{th}}$ moment of the expected distribution of distances between a vacancy "a" seen in area j and the next vacancy seen in the search sequence is

$$m_i = \int_0^\infty s^i \sum_k g(s, k)\, q_{jk}/n_k\ ds$$

where g(s, k) is the number of vacancies in $n_k$ which are a distance s from "a".

(7) The expected number of vacancies seen at a distance $s \leq s^*$ from a reference point, "a", (the household's current residence or the centroid of the search pattern for example) is

$$E(g(s)) = \sum_k \Pi\,\ell_{\cdot k}\, g(s^*, k)/n_k$$

and the second moment of the distribution is

$$E(g^2(s)) = \sum_k M(f^2(k))\ (g(s^*, k)/n_k)^2$$

where $g(s^*, k)$ is the number of vacancies in $n_k$ which are a distance $s \leq s^*$ from a.

(8) The $i^{\underline{th}}$ moment of the expected distribution of distances between vacancies seen by a household during its search and an anchor point "a" is

$$m_i = \int_0^\infty s^i \sum_k K\,\Pi\,\ell_{\cdot k}\, g(s, k)/n_k\ ds.$$

For this particular model, the distribution described above is identical to the distribution of distances between the vacancy ultimately chosen by the household and the reference point, "a". The distance characteristics of the expected search pattern arising from the specific parameter values are displayed in Table 5.1.

## CONCLUDING REMARKS

The residential search model proposed in this paper represents an effort to take what we know or suspect about the spatial aspects of residential search and to express these ideas in formal terms. Inferences pertaining to the search patterns generated by these models are derived thus providing a means of directly comparing the outcomes of differing conceptualizations of the search process. These inferences also provide a mechanism for discriminating between the various models since they can be judged against observed search behavior. The appeal of the expected search cost model is that the interlocking assumptions provide a means of successively uncovering several levels of explanation in answer to the basic question of why the probability of search varies from one area to another. The end result is a search model which is shown to be a first order Markov chain. The probability of searching in a given area is a function of the location of the last vacancy seen by the household and the relative concentration of possible acceptable vacancies. The parameters in the probability density function are all interpreted within the structure of the model.

The inferences derived from the search cost model pertain to the spatial characteristics of the search pattern generated by the model. The number of areas searched and the number of vacancies seen in each area are predicted as well as the distribution of distances between a fixed reference point and the vacancies in the search sequence.

Two of the most restrictive assumptions in the search model are that the probability of success, K, is the same for all areas and the cost of searching in a new area is proportional to the expected cost of searching in that area. In both instances, the model may be modified to allow a broader range of alternatives with an attendant increase in the number of parameters which must be estimated.

The original choice set model does not preclude the possibility that the probability of success varies from one area to another. In fact if the conditions on N are relatively weak as would be the case if S provides very little relevant information on vacancies in H then the probability of success will tend to vary from area to area as a function of the distribution of acceptable vacancies which are likely to be identified by the source S. The model can be easily modified to accommodate the possibility that the probability of success varies across areas.

The second modification stems from the observation that the model in its present form says nothing about the role of the household's prior familiarity with an area in structuring the search process. All else being equal a household will be more likely to search in a familiar area than in an unfamiliar area. In the terms used in the model, the start-up costs are likely to be higher in unfamiliar areas. The set of areas can be dicotomized into familiar and unfamiliar areas with an additional start-up cost attached to search in unfamiliar areas.

## REFERENCES

Barrett, F. (1973) Residential Search Behavior. Toronto: York University Research Monographs.

Brown, L. and J. Holmes (1971) "Search behavior in an intraurban migration context", Environment and Planning 307-326.

Burke, M. G. Belch, R. Lutz, and J. Bettman (1979) "Affirmative disclosure in home purchasing", Journal of Consumer Affairs, 13, 297-310.

Clark, W. A. V. and T. R. Smith (1979) "Modeling information in a spatial context", Annals of the Association of American Geograpahers, 69, 575-88.

Courant, P. (1978) "Racial prejudice in a search model of the urban housing market", Journal of Urban Economics, 5, 329-45.

Cronin, J. (1979) Low Income Households' Search for Housing: Preliminary Findings on Racial Differences. Washington, D.C.: The Urban Institute.

Flowerdew, R. (1976) "Search strategies and stopping rules in residential mobility", Transactions, Institute of British Geographers, 1, 47-57.

Goodman, J. L. (1980) "Information, uncertainty, and the microeconomic model of migration decision making", in G. DeJong and R. Gardner (eds.) Migration Decision Making. New York: Pergamon Press.

Hempel, D. J. (1969) "Search behavior and information utilization in the home buying process", Proceedings of the American Marketing Association, 30, 241-49.

Howard, J. A. (1977) Consumer Behavior: Application of Theory. New York: McGraw Hill.

Kemeny, J., and J. Snell (1960) Finite Markov Chains. Princeton, New Jersey: D. Van Nostrand Co.

McCarthy, K. (1979) "Housing search and residential mobility", Paper presented at the conference on the Housing Choices of Low Income Families, Washington, D.C.

Meyer, R. (1980) "A descriptive model of constrained residential search", Geographical Analysis, 12, 21-32.

Michelson, W. (1977) Environmental Choice, Human Behavior, and Residential Satisfaction. New York: Oxford University Press.

Palm, R. (1976) "The role of real estate agents as information mediators in two American cities", Geografiska Annaler, 58B, 28-41.

Rossi, P. (1955) Why Families Move. Glencoe: The Free Press.

Schneider, C. H. (1975) "Models of space searching in urban areas", Geographical Analysis, 7, 173-85.

Smith, T., W. A. V. Clark, J. O. Huff, and P. Shapiro (1979) "A decision-making and search model for intraurban migration", Geographical Analysis, 11, 1-22.

----- and W. A. V. Clark (1980) Housing market search: information constraints and efficiency", in W. A. V. Clark and E. G. Moore Residential Mobility and Public Policy. Beverly Hills, California: Sage Publications.

Speare, A., S. Goldstein, and W. Frey (1975) Residential Mobility and Metropolitan Change. Cambridge, MA.: Balinger.

Stigler, G. (1961) "The economics of information", Journal of Political Economy, 69, 213-25.

----- (1962) "Information in the labor market", Journal of Political Economy, 70, 94-105.

Tversky, A. (1972) "Elimination by aspects: A theory of choice", Psychological Review, 79, 281-99.

Weinberg, D., R. Atkinson, A. Vidal, J. Wallace, and G. Weisbrod (1977) Locational Choice Part I: Search and Mobility. Cambridge, M.A.: Abt Associates.

Wolpert, J. (1965) "Behavioral aspects of the decision to migrate", Papers of the Regional Science Association, 15, 159-69.

Yinger, J. (1978) "The Black-White price differential in housing: some further evidence", Land Economics, 54, 187-206.

# Part II

# INFORMATION, SEARCH, AND POLICY RESPONSES

The second part of the book is concerned with the nature and role of information as it affects the search process and with the policy implications of studies of residential search. Because the policy implications are often related to the provision of information, as is clearly demonstrated in at least two of the chapters, there seems to be a natural connection amongst search, information, and policy. The chapters include studies of information acquisition (the demand side), information provision (the supply of information, especially the interrelationship of real estate agents and their use of the newspaper as an advertising medium), buyer response to information content, and the impact of information on the links between migration and subsequent residential mobility. The book concludes with an essay which evaluates the relative role of supply and demand approaches to understanding residential search.

Chapter six in this second section of the book lays out the important role of information as a central element of housing market search. As MacLennan and Wood note, "a successful search for housing is not a single activity and can be more appropriately viewed as consisting of a linked series of distinctive information seeking actions". They outline a broad framework in which issues related to information network choice can be assessed and set up a simple model which is used to develop some specific hypotheses. They test these hypotheses with data for renters from the controlled rental market in Glasgow and for first-time home buyers in the same market. In both cases, they focus on the nature of channel choice and the way in which these channels are used to establish vacancies and to aid the acquisition of quality housing. Their major conclusion is that an information acquisition/information use approach considerably enhances understanding housing choice processes and outcomes.

In analyzing the impact of information on residential decision-making, and specifically, on household relocation, it is important to distinguish between the demand side which is the process of information acquisition by individuals in the housing market and the supply side (realtor information on vacancies for example). The chapter by Smith, Clark, and Onaka analyzes information provision, specifically information available from newspaper advertisements and the role of agents as information transmitters. Using data from two housing markets in California for a twenty year period, the chapter provides some of the first evidence of the importance of supply side analyses of information provision. Although the chapter reports quite preliminary results, the research clearly shows that there are different patterns of information provision by owners and real estate agents and for the different markets over time. The major hypotheses are focused on the nature of ambiguity in advertising and especially contextual (market) influences on ambiguity. Although the results are tentative, they do suggest that owners are less certain about price distributions and provide less information, that more active markets have less certain price information, and that there is a tendency for real estate agents to treat price and location as partial substitutes to create a maximum level of ambiguity.

The increasing impacts on house sales transactions of both local and federal legislation can be seen as part of the impetus towards consumer protection and specifically as part of the role of government in information provision as a mechanism for consumer protection. Risa Palm focuses on the mandated provision of information on natural hazards and the role of the real estate agent as an informational agent in this process. The results are not encouraging. Individuals place more emphasis on housing as an investment than as shelter, and consequently, buyers do not consider that information acquisition on issues other than location, price, and house characteristics is important. And, real estate agents are poor information channels for this natural hazard information.

Information is also a critical variable in the relationship between long distance migration and later local housing adjustments (mobility). While we know that long distance migration is primarily motivated by economic considerations and that local mobility is influenced by housing adjustment/life cycle explanations, the role of information in linking them has not been adequately specified. One hypothesis -- the same-person hypothesis -- suggests that because there is less information about the destination area available to long distance migrants, the long distance migrants are more likely to make subsequent adjustment moves. A second hypothesis -- the place effect hypothesis -- suggests that high rates of in-migration affect the local mobility rates of current residents by altering neighborhood composition and housing market conditions. Goodman examines the first of these two hypotheses as part of a larger analysis of variations in local mobility rates and considers the implicit role of information in residential buyer behavior. The paper shows that there is a significantly higher probability of long distance migrants selecting temporary housing in the new location.

The final chapter evaluates the policy implications of studies of search behavior. As Eric Moore points out, the standard approach in the search literature is to analyze the consumer demand for information as he moves through the housing market. Much of the literature is focused on differences amongst groups (particularly minorities) in terms of the search procedures and strategies that they employ in the housing market. But from a policy perspective, the nature of the outcomes (of search) may be much more important. What is the difference in the outcomes for households who move and for those who do not move? What is the role of the local context and its impact on search? And, what is the role of the supply of information in influencing outcomes of search? The chapter attempts to place the roles and strategies of search in the housing market into a broader policy perspective. Moore argues quite convincingly for greater attention to the behavior of supply side actors in the explanations of search behavior, especially if the concern is to develop intervention strategies for improving housing consumption. The argument is clearly one for further investigations of housing market search.

In sum, the combination of careful analytic modelling, empirical testing, and speculative discussions of the role of agents in the search process provides an innovative and imaginative approach to the interesting problems of understanding individual behavior in complex contexts such as the urban housing market.

# 6

# INFORMATION ACQUISITION: PATTERNS AND STRATEGIES

**Duncan Maclennan and Gavin Wood**

The processes of information acquisition, processing and utilisation have come to be of theoretical and empirical concern in a range of social science disciplines. Empirical investigations in psychology, studies of geographic search and marketing research focusing on information acquisition and utilisation have now become commonplace. However, economic studies incorporating information acquisition have largely been concerned with theoretical questions rather than empirical analysis. And, the bulk of empirical and theoretical studies concerned with information acquisition have been undertaken in relation to labour market analysis. In this chapter we wish to extend, but with an empirical rather than theoretical orientation, economic analysis of information gathering to the housing market.[1] In particular we wish to stress that housing market information gathering and search will almost always be an important aspect of housing choice behaviour (see also Wood and Maclennan, chapter 3 of this volume for additional comments on this issue). It will also be suggested that housing market search is not completely random and that households rationally select and re-select information channels depending on their initial information and subsequent experience. Moreover, it is argued that the direct and indirect costs of information acquisition are important in the housing system and that, in certain circumstances, institutional "experts" in the housing market may manipulate information flows to satisfy their own objectives at the cost of customer interests.

## THE HOUSING SEARCH AND PURCHASE PROCESS

A successful search for housing is not a single activity but can be more appropriately viewed as consisting of a linked series of distinctive information seeking actions (this will be true of most commodities). Therefore studies of housing and labour markets which merely seek to identify, expost, the source of information by which a house or job are found omit two important considerations. First, in any search stage information use may alter and, second, information channel choice may vary at different search stages. In the housing


DOI: 10.4324/9781003182085-8

search process we hypothesize that initially the household chooses to actively seek new housing, then a series of information channels are selected. Next a broad area neighbourhood orientation phase occurs, then housing vacancies have to be established. In turn, detailed evaluation of identified vacancies may then be required prior to a price bid or contract being offered. Clearly the capacity of individuals to undertake such search tasks may vary across discernible groups and the sources and types of information required may also alter from stage to stage.[2] Let us consider this elaboration of the "mover-stayer" model in more detail along a chain of search stages as indicated in Figure 1.

## Move - Stay Decisions

The individual household has some initial combination of housing services and housing payments. Active housing search can be precipitated by a series of factors. Sharp and sudden events such as family bereavement, or job loss can "push" the household into market search. Alternatively, a series of longer term gradual stimuli may "pull" the household into the market. Even at this stage of search, the information influences on "push" and "pull" motives are likely to be different. "Push" messages are likely to be immediate, personal and direct. "Pull" messages may emerge from memory or a "concept learned" word or from a steady flow of subliminal or non-specific housing information (Maclennan 1977, Maclennan 1981).

## Channel Selection

Whether "pull" or "push" factors first trigger search effort, from the economic standpoint, it can be argued that housing search will take place if the household believes that expected alternative housing opportunities and their associated rents or prices will be superior to present housing circumstances after allowing for expected search and transaction costs. All individuals, on setting out on the housing search process, are subject to some uncertainty about the nature of information services and housing price/quality/availability options. Thus household search decisions are beset with uncertainty. In a recent survey article, Hirshleifer and Riley (1979) made a distinction between terminal and non-terminal decisions in the face of uncertainty. Terminal decisions reflected situations where individuals did not have perfect knowledge regarding the outcomes of actions and were not able to improve their information. Non-terminal decisions represented situations wherein actions could be revised in the light of outcomes - that is, learning could reduce uncertainty.

At the initiation of search not all households are identical. They may differ in their past familiarity with an area and in their past use of the housing market. Further, they may have different incomes, which will directly affect search strategies where costly information services exist.[3] Finally, they may be more or less risk

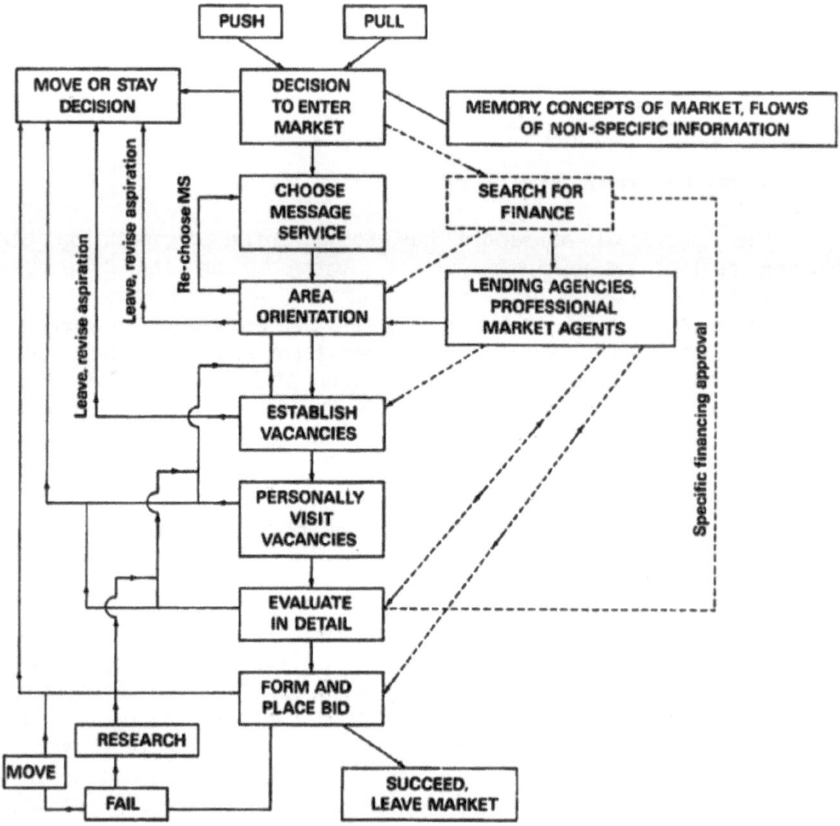


Figure 6.1. The chain of search stages.

averse, so for instance, some households may be prepared to risk longer search duration by initially utilising low cost but low efficiency message services. Having chosen a message service(s) the household is unlikely to immediately revise their sampling strategy. It is only after a number of uses of the service(s) that households may realise that initial confidence in the messenger service had been misplaced.[4] Thus at any stage of the search process we are interested in establishing whether households with different experience, incomes, etc. select different message services and whether these selections are "non-terminal", that is whether sources used will be revised or extended. In the following section specific hypotheses of this kind are formulated and tested in relation to our renter sample.

## Area Orientation

The initial message service selected for the household will be directed towards the identification of broad housing sector opportunities. This may be labelled, following management science literature, the consumer "orientation" phase. That is, the searcher establishes his broad preferences for house type location etc. and also identifies the areas or sectors of the local housing market to which he may, given perceived constraints, realistically aspire. Specific household aspirations may focus on a number of housing attributes. Size or style may dominate location as a choice consideration. But, in the urban context examined here, the pre-eminence of spatial orientation is not an unreasonable assumption - at the very least, size, style, age and location are all highly correlated. This search phase, which is relatively costless and quickly undertaken, is akin to "research and development" activities. That is broadly defined opportunities for housing consumption are identified with a view to more detailed investigation. And this search phase is highly rational as it leads to the avoidance of detailed costly searches in areas which are broadly unacceptable.

In this area search phase we would not expect households to employ "professional expertise". For not only is the establishment of preferences involved, but area quality can be reasonably quickly established by personal visits. In this instance, we would expect local searchers to utilise personal knowledge and the advice of friends and neighbours, etc. Non-local searchers may use professional agents and direct visits. But "formal" information sources are not likely to be paramount at this stage of search. A major qualification to this argument, as detailed in the section on information sources for first time home buyers, is where potential house purchasers are also seeking finance as well as housing. In these instances institutions, used as a source of advice, may structure or constrain the "orientation" process.

## Information acquisition

### Identifying Vacancies

Having successfully "orientated" their housing market search activity, the individual then has to establish the existence of specific purchase or rental opportunities. As is indicated in the next section, the importance of this phase will clearly depend on the degree of excess demand and the nature of the market system. Where there is excess supply a relatively leisurely perusal of formally advertised vacancies may be appropriate. However where excess demand prevails the establishment of unfilled vacancies is critical, particularly if there is non-price rationing and a first-come first-served allocation principle by sellers. In such instances, which we believe will characterise contexts of effective rent controls, the information acquisition strategy becomes crucial and the desirability and cost of acquiring market information is shifted to the potential consumer.[5] In this search phase we would anticipate that not only are "informal contacts" relatively highly valued and effective, insofar as they transmit information to a relatively small number of searchers but all groups will also use formal sources. The formal information channel which disseminates information most widely will differ across markets.

### Assessing Vacancies

When the household has established the existence of a vacancy in a desired neighbourhood further more intensive research may be required. In our renter sample examined later in this chapter, there existed a narrow range of quality options with non-price rationing accompanying excess demand, and the identification of an existing vacancy was a general search "stopping" criterion. But in the owner sample, the household required further information related in particular to the durability and asset aspects of housing. First, because most houses are technically complex structures in which sellers could disguise or conceal the existence of long-term and costly defects, the household will supplement their own information gleaned from visiting the property by paying for the services of a professional evaluator or surveyor. Further, institutions lending for house purchase will usually require a survey to recommend purchase and they will require the surveyor to be trustworthy or of a known, approved reputation. In the British context, since building societies dominate lending for house purchase it is they who instruct surveyors to examine houses on which searchers have asked for a loan. And although the searcher pays for the survey, most societies reserve the right to restrict the distribution of information contained within the report. A potential conflict of consumer interest with lending institution goals may arise at this stage. Since surveys are a relatively expensive means of acquiring property specific information, we examine in the empirical sections of this paper the reasons why some surveys do not proceed to the bid formation stage.

Bid Formation

Even when a suitable vacancy has been identified and detailed assessment indicates that it is suitable for purchase and loan finance, the consumer's information search process has not yet finished. Indeed, we have only now reached the stage at which the Walrasian auction of economic theory begins. The individual has to make a price bid. In an open auction system bidding would be a non-terminal decision in the sense that if a bid failed it could be revised, as the auction proceeded, without the incurrence of any additional transaction or search costs to the potential buyer. But not all housing markets are open auctions or even close approximations thereto. In the Scottish system, house sales are generally enacted by sealed bids being placed with a seller's agent. Although a reservation price is known, the searcher may not know how many bidders are involved nor is he necessarily aware of the prices of similar houses sold in the same market period. In placing the price bid, therefore, a decision of a "terminal" kind is made in relation to that property. Of course in forming the bid in relation to the reservation price the searcher may discount future potential search costs into the bid or, if several failed bids have been made, they may begin to identify the required successful margin in excess of a particular reservation price. This indirect bidding system not only fails to immediately reveal prices to subsequent purchasers, but most crucially does not allow searchers to revise bids. That is, the price information in the system is imprecise and out of date and, as a result, costly non-price information seeking strategies are required. At this bid formation stage we would expect professional message services to be significant sources of information.

Those comments have suggested a broad, general framework in which issues related to information network choice may be assessed. From two "search process" focused housing market studies we are able to report some specific hypotheses and tests in relation to parts of this suggested framework. With respect to the renter sample, we explore specific hypotheses regarding influences on channel choice and revision of channel choice in the post orientation phases of establishing vacancies. For the owner sample, apart from identifying influences on channel choice, it is possible to indicate how channel choice varies at different search stages and how institutional guidance modifies the search process.

## INFORMATION ACQUISITION IN A CONTROLLED RENTAL MARKET

The first series of tests reported here relate to a detailed study of student and non-student search for furnished rental housing in the City of Glasgow in the period 1974-1976. The legislative nature of controls and their supplyside impacts are discussed elsewhere (Maclennan 1978).

Information acquisition

Searcher reactions to excess demand for rental housing have also been reported (Maclennan 1979a, 1979b, 1981b). The details are not reiterated here but some preliminary points should be stressed. A detailed survey of actual search processes indicated that rental housing searchers did use a variety of information sources, informal channels were important and the overall pattern of channels used differed significantly from the structure of successful channels (Maclennan, 1979b). Further, it was established that information sources did vary in their frequency of transmission, costs of utilisation, success of placement rates and in the quality of the housing vacancies advertised. These findings merely emphasised the necessity of monitoring and analysing the total search process and of explaining channel choice and use. It is to this latter issue we now turn.

A Model of Channel Choice

For the purposes of simplication let there be two information channels, t and m, where these variables indicate volume of message use. We assume that channels t and m are used for two purposes.

(1) to establish vacancies
(2) to aid the acquisition of quality housing

There is no reason to believe both channels to be equally efficient with regards to these two criteria. We therefore make use of a joint probability distribution defined with regard to t and m:

$$\hat{b} = \hat{b}(t, m) \qquad \partial b/\partial \hat{t} > 0, \qquad \partial \hat{b}/\partial m > 0. \tag{1}$$

Where $\hat{b}$ is the probability of establishing a vacancy from the use of t, m combinations. Additionally, expected housing quality is linked to channel use:

$$\hat{H}_s = \hat{H}_s(t, m) \qquad \partial \hat{H}_s/\partial t > 0, \qquad \partial \hat{H}_s/\partial m > 0. \tag{2}$$

Where $\hat{H}_s$ can be measured in terms of expected quantity of housing services.[5] In order to utilise the framework of constrained utility maximisation we ascribe to the representative individual a utility function:

$$U = U(\hat{H}_s, \hat{b}, C) \quad \partial U/\partial \hat{H}_s > 0, \partial U/\partial \hat{b} > 0, \partial U/\partial C > 0. \tag{3}$$

Where the arguments are $\hat{H}_s$, $\hat{b}$ and C, a composite commodity representing all other goods (relative prices assumed constant), and the individual is presumed to be risk averse. Upon substituting (1) and (2) into (3) we can define

$$U = U'(t, m, C) \tag{4}$$

Channel use is costly, so let $g$ and $\delta$ be the unit price of using t and m respectively. Further, $p_h$ and $\epsilon$ denote the prices of $H_s$ and C respectively. The Budget constraint can then be defined as

$$Y = p_h \hat{H}_s + gt + \delta m + \epsilon C. \tag{5}$$

The first order conditions for utility maximisation are

$$\frac{\partial U'}{\partial t} + \lambda(-p_h\, \partial\hat{H}/\partial t - g) = 0 \tag{6}$$

$$\frac{\partial U'}{\partial m} + \lambda(-p_h\, \partial\hat{H}/\partial m - \delta) = 0 \tag{7}$$

$$\frac{\partial U'}{\partial m} - \lambda\epsilon c = 0 \tag{8}$$

$$Y - p_h\hat{H}_s - gt - \delta m - \epsilon C = 0 \tag{9}$$

Dividing 6 by 7 we have the familiar condition

$$\frac{\partial U'/\partial t}{\partial U'/\partial m} = \frac{p_h\ \partial\hat{H}_s/\partial t + g}{p_h\ \partial\hat{H}_s/\partial m + \delta} \tag{10}$$

that rational channel choice requires satisfaction of the condition that the marginal rate of substitution be equal to the ratio of the marginal costs of channel use. From (1), (2) and (3) we can rewrite 6 and 7:

$$\frac{\partial U}{\partial \hat{H}_s}\frac{\partial \hat{H}_s}{\partial t} + \frac{\partial U}{\partial \hat{b}}\frac{\partial \hat{b}}{\partial t} + \lambda\left(-p_h \frac{\partial \hat{H}_s}{\partial t} - g\right) = 0 \tag{11}$$

$$\frac{\partial U}{\partial \hat{H}_s}\frac{\partial \hat{H}_s}{\partial m} + \frac{\partial U}{\partial \hat{b}}\frac{\partial \hat{b}}{\partial m} + \lambda\left(-p_h \frac{\partial \hat{H}_s}{\partial m} - \delta\right) = 0 \tag{12}$$

Use of 11 and 12 allows 10 to be rewritten as (13).

$$\frac{\dfrac{\partial U}{\partial \hat{H}_s}\dfrac{\partial \hat{H}_s}{\partial t} + \dfrac{\partial U}{\partial \hat{b}}\dfrac{\partial \hat{b}}{\partial t}}{\dfrac{\partial U}{\partial \hat{H}_s}\dfrac{\partial \hat{H}_s}{\partial m} + \dfrac{\partial U}{\partial \hat{b}}\dfrac{\partial \hat{b}}{\partial m}} = \frac{p_h \dfrac{\partial \hat{H}_s}{t} + g}{p_h \dfrac{\hat{H}_s}{m} + \delta} \tag{13}$$

It is clear from the first order conditions that both the marginal cost of channel use and reliability of information channels are important influences on choice. There are two important comparative static considerations, first the overall level of utilisation of t, m combinations, and secondly, the pattern of use.

If we treat t and m as a composite good (service) S, then it is possible to treat the comparative statics of overall level of utilisation by means of traditional income and substitution effect analysis. Figure 6.2 depicts situation in which $g$ and $\delta$ increase by an equi-proportionate amount. Clearly, the overall direction of change depends upon the nature of the income effect. If this is in the opposite direction to the substitution effect and outweighs the latter, then S is a Giffen good as is illustrated by equilibrium choice C′with an increased utilisation of S. The conventional case of a normal good is depicted by equilibrium choice C with a lower utilisation of S.

Ambiguity also characterises the effect of change in the relative prices of t and m. Further, given the presence of C, the comparative statistics are complicated by whether or not t and m are gross substitutes or gross complements. The full range of possible responses to change in $g$ and $\delta$ are depicted in Table 6.1. The general expectation would be that of treating t and m as normal goods and gross substitutes. Hence, with regard to a fall in the price of the m channel, the utilisation of t would fall while that of m increases.

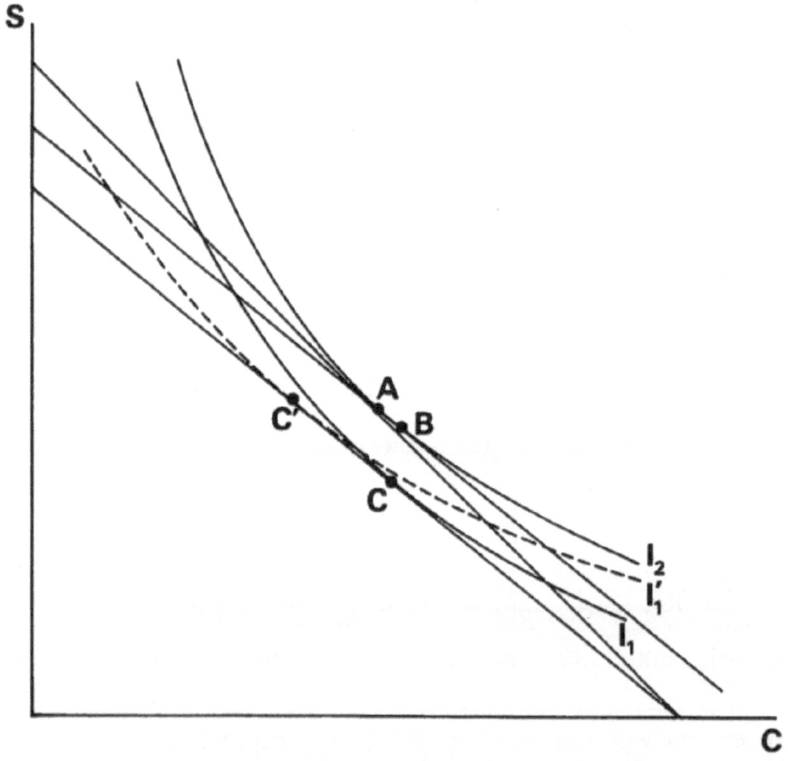


Figure 6.2. Comparative statics of channel utilization.

However, as Table 6.1 indicates, it is feasible for t and m to both be Giffen goods but to also be gross complements in which case the fall in the price of the m channel would lead to a decline in the utilization of both channels.

Changes in $g$ and $\delta$ are one influence upon channel choice. The model specification also emphasises the importance of the reliability of the information channels. The influence of reliability is exerted by its direct impact upon the individual's preference ordering. Clearly, from the specifications of (1), (2) and (3) above, an improvement in a channel's contribution to probability of acquisition will positively influence preference orderings with regard to that channel. Hence, ceteris paribus, we can unambiguously expect an increased volume of use. Improvement in a channel's expected quantity of housing services will have an ambiguous influence as a consequence of the higher expenditure this entails.

Specific Hypotheses

This simple model of channel choice provides a number of hypotheses which are in principle testable. Any sample of individuals is likely to exhibit variations in the marginal costs of channel use, which are attributable to differences in housing market experience and the degree of excess demand prevailing during the search period. However, empirical examination is complicated in the current context for three reasons. Firstly, a substantial proportion of the sample of movers were already tenants in the market for whom

Table 6.1: Possible Responses to Changes in $g$ and $\delta$

| | Gross Substitute | Gross Complement | Gross Substitute | Gross Complement |
|---|---|---|---|---|
| Normal Good (t) | - | - | + | - |
| Giffen Good (t) | + | + | + | - |
| Normal Good (m) | + | - | - | - |
| Giffen Good (m) | + | - | + | + |

channel use was therefore conditional upon whether it would ensure a position preferable to that already attained. Secondly, those who are students in the sample face a time constraint which, when binding, will ensure a choice which differs from that which would otherwise prevail. Lastly, the sample data does not ascertain quantitative measures of the costs of channel use and expected reliability of information channels. It is not therefore possible to undertake rigorous tests of the predictions afforded by the simple model of the previous section. We therefore undertake the less ambitious task of partitioning the sample and investigating differences in subgroup patterns of channel choice in accord with implications of our model. The following investigations are suggested by our discussion.

(1) Though all members of the student subgroup are moving into rather than within the sector, they will differ with regard to their past experience in the local housing market. In particular, newcomers are likely to experience a higher relative price in using informal information channels compared to students with past experience now re-searching. Ceteris paribus, newcomers will use formal channels more intensively if categorised as a gross substitute.

(2) For the non-student group, movers can be classified into the subgroups of newcomers/re-entrants, and residents whose previous address was in the sector. For the latter group $\partial\hat{H}_s/\partial t$, $\partial\hat{H}_s/\partial m$ are only relevant considerations when channel use provides opportunities to improve upon present housing satisfaction in the sector. Additionally, as their experience in the sector will be greater than that of newcomers/re-entrants, they will in general face a lower relative price in using informal information channels. Thus, intra-sector movers will undertake a lower overall utilisation of information channels and, subject to the gross substitute condition, a lower number of formal sources than newcomers/re-entrants.

(3) Where excess demand in the sector varies across time periods we can expect acquisition probabilities and prices of channel use to vary correspondingly. Acquisition probabilities will deteriorate with increases in excess demand; this factor will exert a contractionary influence upon overall channel use. The prices of information channels can be expected to increase with positive variations in excess demand, and subject to the normal good condition, will also lower overall channel use.

## The Results

<u>Market experience</u>. Table 6.2 indicates that local and more experienced students use sets of information sources which are, at

the 0.05 significance level, distinctly different from those utilised by new and non-local students. As we anticipated, local and experienced students have a much higher proportion and number of informal contacts in their information supply. Local students were less likely to use official agencies, and though both groups utilised newspapers extensively, personal contacts and shop windows were more important for experienced students. It is also noteworthy that new students undertook a lower overall level of utilisation of information channels. It is possible that term start dates represent a more restrictive constraint on search duration for new students than experienced students, and this is then reflected in a lower level of utilisation.

Continuing non-students versus newcomers and re-entrants. The hypotheses are confirmed by the data presented in Table 6.3 . On average individuals moving within the sector undertook less than five searchers, with 64 percent of the messages arising from friends or relatives. The newcomers and re-entrants undertook almost twice the number of searches mostly based on formal sources. In part, since there was a tendency for continuing tenants to search outside the peak period, these results are influenced by the hypothesis reported immediately below. Therefore the results in Table 6.3 control for this effect by reporting the behaviour of continuing and newly-entering groups for the non-peak period.

Table 6.2: Information Source by Experience Status

| | Continuing | | Newcomers | |
|---|---|---|---|---|
| | Number | Relative frequency | Number | Relative frequency |
| Lodgings office | 1.52 | 10.83 | 1.99 | 17.97* |
| Newspapers | 4.00 | 28.51 | 4.33 | 39.79* |
| Shop window | 4.31 | 31.21 | 0.96 | 8.67* |
| Estate agent | 2.14 | 16.97 | 3.53 | 31.11* |
| Friends, relatives | 1.78 | 12.87 | 0.27 | 2.43* |
| Landlords | 0.01 | 0.07 | ---- | ---- |
| Total | 14.03 | | 11.07 | |
| | N = 732 | | N = 269 | |

*Significantly different from continuing group mean at the 0.05 level.

Information acquisition

The excess demand effect. We can establish the importance of this market tightness effect for non-students since, unlike students, they search through out the year. Table 6.4 indicates that there are significant differences in information structures between peak/non-peak periods. In the peak period non-students not only have a higher overall utilisation of channels but they also utilise a greater number or higher proportion of formal channels than searchers outside the peak. So it would appear that excess demand is an important influence but of a nature counter to that anticipated. Searchers react to the higher costs of channel use and deteriorating probability of acquisition by increasing their overall use of channels. Though this is not ruled out by the simple model presented earlier, it does suggest that these individuals were unwilling in response to higher costs of channel use to trade-off risk of acquisition for higher consumption of other goods. Given the importance of the necessity elements of housing this is perhaps not surprising.

Table 6.3: Information Acquisition for the Non-student Sample: Searchers Outside Peak

| | Entrants (N = 78) | | Continuing (N = 162) | |
|---|---|---|---|---|
| | Number | % | Number | % |
| Friends in house | 0.3 | 3.57 | 1.63 | 34.53* |
| Friends in area | 0.7 | 8.33 | 1.79 | 37.36* |
| Newspapers | 3.9 | 46.42 | 0.58 | 12.17* |
| Card | 2.8 | 33.33 | 0.73 | 15.40* |
| Estate agent | 0.7 | 8.33 | 0.05 | 0.01* |
| Other landlord | --- | --- | --- | --- |
| Total | 8.4 | 100.00 | 4.75 | |

*Significantly different from new entrants at the 0.05 level

Table 6.4: Information Acquisition for the Non-student Sample, Newcomers Excluded

| | Peak sources | | Non-peak | |
|---|---|---|---|---|
| | Number | % | Number | % |
| Friends already in the house | 0.70 | 7.50 | 1.63 | 34.53 |
| Friends in area | 1.40 | 15.00 | 1.79 | 37.76 |
| Newspapers | 3.20 | 34.40 | 0.58 | 12.17 |
| Card in shop window | 2.80 | 30.10 | 0.73 | 15.40 |
| Estate/rental agent | 1.18 | 12.69 | 0.05 | 0.01 |
| Other landlord | 0.02 | 0.20 | ---- | ---- |
| Total | 9.30 | | 4.75 | 0.01 |
| | N = 42 | | N = 162 | |

*Significantly different from peak sources at the 0.05 level

Search stage. In an earlier paper it was suggested that the search strategies of students were likely to be revised as the search process produced a revision of the expected sectoral pattern of vacancies (Maclennan 1979b). Though not considered by the static model elaborated earlier, a similar revision of information source use is likely to occur as individuals not only ascertain mode efficiency but also revise general search expectations. In general it could be anticipated that newcomers to the market would initially use a wide range of formal networks and only if search duration were extended would friends or personal contacts become an information source. We therefore hypothesize that the proportion of "informal" based searches rises for newcomers as the duration of search increases. It is more difficult to specify the likely changes for experienced market searchers. An initial dependence on informal networks with increasing formal usage on search extension is a plausible hypothesis. Presumably search is extended because vacancies are not forthcoming within informal channels and, therefore, the individual with a limited search time horizon has to spill over into formal networks.

Since search records are only available for the student search record sample the results of this analysis are restricted to students only. The sub-populations identified are experienced students, that is students returning to the local housing market or local students entering student housing for the first time, versus inexperienced students. The latter group were students who were entering the city housing market for the first time and who did not have a local home address. The results are based on information sources for four successive units of search. The reported statistics in Table 6.5 confirm the hypothesis for inexperienced students. There is a progressive extension of their use of informal networks throughout the search process. However, the results for experienced students do not confirm our initial hypothesis. Instead, the proportion of formal contacts (lodgings office) used is high for early searches, falls in the middle phase and rises towards higher number of searches. Within the constraints of the data available, it is not possible to explain this phenomenon. However, it is plausible that experienced students first survey a wide sample of vacancy sources, then relearn or begin to take account of channel efficiency and then latterly revert to a variety of sources as their informal contacts fail.

It is important however, to realize that this is only one of several possible ex post rationalisations of the observed results. It should serve no purpose other than as a reminder that search studies should ascertain the reasons for channel choice at each stage of the search sequence. Table 6.5 also indicates that the information sources of newcomers and continuing students are substantially different in the early search phases. In the final phase of search the information source structures are remarkably similar, presumably reflecting a learning process on the part of newcomers.

The empirical tests largely confirm that our a priori notions of influence on channel choice are not unreasonable. However, in this example we were primarily concerned with information acquisition at the vacancy establishment stage and there was no major institutional involvement. The second example emphasises changes in channel use with search stage and the importance of institutions.

## INFORMATION SOURCES FOR FIRST-TIME HOME BUYERS

The data context for this section was a detailed survey of 524 first-time home buyers operating in the Glasgow housing market in 1977. This survey of household socio-economic characteristics, housing chosen and housing and finance search processes covered a one in three sample of new entrants in the relevant time period.

An analysis of channel choice confirms the significance of factors important in the rental sample. Although the new entrant sample were all making a house purchase decision for the first time, and were therefore unfamiliar with the housing purchase process, backgrounds of the sample varied markedly and this was reflected in channel choice. For instance, non-local orginating entrants initially used more formal channels, as did entrants who did not have a parental background in owned housing. Further, higher income groups were more likely to use estate agents and mortgage consultants as information sources. Households "pushed" rather than "pulled" into market activity did use, initially, a wider range of information sources reflecting greater temporal urgence of search. However, in

Table 6.5: Relative Frequency Distribution of Information Source Use by Stage of Search for New and Continuing Student Groups

| | Continuing | | | | Newcomers | | | |
|---|---|---|---|---|---|---|---|---|
| | 1 to 4 | 5 to 8 | 9 to 12 | 13 to 16 | 1 to 4 | 5 to 8 | 9 to 12 | 13 to 16 |
| Friends/ relatives | 7.1 | 19.8 | 16.9 | 6.3 | 0.2 | 1.4 | 3.5 | 5.4 |
| Other landlords | --- | --- | --- | 0.3 | --- | --- | --- | --- |
| Lodgings office | 13.2 | 6.7 | 8.7 | 15.5 | 26.7 | 24.2 | 15.3 | 16.1 |
| Shop window | 30.6 | 31.0 | 29.8 | 38.4 | 2.3 | 10.7 | 24.5 | 31.2 |
| Newspapers | 42.1 | 30.3 | 31.7 | 24.3 | 47.1 | 42.3 | 37.7 | 29.0 |
| Estate agents | 7.0 | 10.2 | 12.9 | 15.2 | 23.3 | 21.4 | 19.0 | 18.3 |

this section our emphasis is not in cross-sample variation in channel choice. Instead, we wish to highlight how channel use changes between search phases.

## The Orientation Phase

In this search phase the household is attempting to pre-structure more detailed search effort by identifying areas which, on first scrutiny, provide house types, neighbourhood characteristics and accessibility broadly in line with aspirations. In addition, since a loan is required for most house purchase the household has to ascertain the extent of their credit-worthiness, any likely delays in having finance available, and they may also seek to establish whether financial institutions have lending rules which relate directly to property or area type. Thus, at least in the Scottish housing system, financial institutions may play a critical role in structuring or orientating the households search to purchase housing assets. These orientation visits tended to be of a "one-time" nature and the kind of information conveyed by institutions to clients is outlined in Table 6.6. Most commonly households visited, in conjunction, a solicitor and a building society. Only higher income groups utilised mortgage consultants and, in general, lower income groups visited local authority home loans offices.[6]

Table 6.6: Institutional Guidance in the Orientation Phase

| Agency | Proportion Visiting Agency | Proportion of those visiting receiving advice on: Loan Amount | Waiting Period | Positive Area | Negative Area |
|---|---|---|---|---|---|
| Building Society | 50.1 | 69.5 | 44.3 | 12.6 | 25.4 |
| Mortgage Consultant | 19.5 | 59.4 | 32.3 | 55.4 | 33.3 |
| Insurance Co. | 19.8 | 49.4 | 22.7 | 10.6 | 17.7 |
| Estate Agent | 30.3 | 23.4 | 11.2 | 11.2 | 11.2 |
| Solicitor | 49.9 | 22.2 | 9.1 | 3.2 | 7.8 |
| Local Authority HL | 34.6 | 64.2 | 22.3 | 2.3 | 4.0 |
| Other | 6.6 | 51.5 | 19.7 | 27.0 | 50.0 |

These institutional visits usually occured simultaneously with "area orientation" searches and had two major effects on search activity. First, the advice regarding likely loan amounts available established a filter for examining broad neighbourhoods or specific vacancies. Second, and this aspect is pursued in more detail elsewhere (Maclennan and Jones, 1981), all institutions, with the exception of local authority home loans, systematically urged households to search outside the inner city area and, more strongly, to avoid the central city. The threat of loan application refusal usually compelled households to comply with this advice or information.

When the relative frequency of sources utilised to establish suitable neighbourhoods for purchase was examined, "informal" information sources are particularly important. Personal knowledge based on regular use of the city and the advice of friends and relatives were markedly more important than in other search phases. However, as indicated in Table 6.7, even at this phase area related information was also gleaned from newspapers and estate agents. Households were asked to indicate, on a scale of 0 to 5, the relative importance of channels in the search phase. Table 6.8 indicates that personal knowledge, newspapers and friends and relatives were the important "orientation" sources of information.

Table 6.7: Information Sources in Different Search Phases

| Source | Selection of Search Area | Establishing Housing Availability | Suggesting New Areas | Found Present House | Overall Most Useful |
|---|---|---|---|---|---|
| Personal Knowledge | 72.6 | 15.0 | 0 | 4.7 | 2.4 |
| Friends & Relatives | 31.4 | 18.2 | 24.0 | 15.1 | 5.6 |
| Newspapers | 34.0 | 76.0 | 64.0 | 55.9 | 58.8 |
| Estate Agents | 17.2 | 37.5 | 4.5 | 12.7 | 9.0 |
| Other | 5.3 | 5.8 | 0.3 | 6.0 | 3.0 |
| Sample Size | 524 | 516 | 109 | 524 | 482 |

Table 6.8: Source of Information for Area Orientation

| Information Channel | Score[a] | | | | | | |
|---|---|---|---|---|---|---|---|
| | 5 | 4 | 3 | 2 | 1 | 0 | mean score |
| Personal Knowledge | 29.3 | 21.4 | 5.8 | 0.8 | 2.1 | 40.6 | 2.53 |
| Friends and Relatives | 2.6 | 13.2 | 8.4 | 2.9 | 1.6 | 71.2 | 0.99 |
| Newspaper Advert. | 7.4 | 12.4 | 5.5 | 3.2 | 2.6 | 68.9 | 1.12 |
| Estate Agents | 1.1 | 5.5 | 4.5 | 2.9 | 1.6 | 84.4 | 0.48 |
| Other | 1.3 | 3.2 | 1.1 | 1.1 | - | 93.4 | 0.25 |

Source of Information on Availability of Housing in Defined Search Areas

| Information Channel | Score[a] | | | | | | |
|---|---|---|---|---|---|---|---|
| | 5 | 4 | 3 | 2 | 1 | 0 | mean score |
| Personal Knowledge | 4.0 | 5.3 | 2.9 | 0.5 | 0.3 | 87.1 | 0.51 |
| Friends and Relatives | 3.4 | 3.7 | 4.0 | 2.6 | 1.1 | 85.2 | 0.50 |
| Newspaper Advert. | 31.9 | 23.0 | 4.2 | 1.6 | 5.8 | 33.5 | 2.73 |
| Estate Agents | 3.7 | 11.6 | 9.5 | 8.7 | 3.2 | 63.3 | 1.14 |
| Other | 1.6 | 3.2 | 1.6 | 1.6 | 0.3 | 91.8 | 0.29 |

*Note:* a. 0 = Not even considered
1 = Not at all important
2 = Not very important
3 = Important
4 = Quite important
5 = Extremely important

Establishing Specific Vacancies

Information regarding property vacancies was most commonly acquired via newspaper adverts and the use of estate agents, although the informal channels had, as in the rental study, higher placement rates. But, as indicated in Table 6.7 for most households the critical channels were formal sources. Tables 6.7 and 6.8 strongly support our contention that the use of information channels will differ across search stages.

Detailed Evaluation Phase

The interview evidence available suggests that most households enacted 2 or 3 explicit orientation visits, and 12 to 16 specific vacancy identification searches by personal visiting. However, not all of the direct visits, where the household's own examination determines decisions of whether or not to terminate search, proceed to surveys. In fact, on average only 20 percent of property visits produced further formal evaluation of that house. Thus, in selecting specific vacancies direct visiting is a critical filter.

The rejection of specific vacancies by direct visiting reflects an inherent, but occasionally deliberate, lack of advertised information by sellers. The reasons why households generally screened out such properties are indicated in Table 6.9. More than 60 percent of search terminations at this stage, reflected searchers negative subjective reactions to specific aspects of the property. But, interestingly, 20 percent of "discards" were for area or price brand reasons which should in principle, have been eliminated by successful orientation and should not have necessitated direct visits.

The detailed evaluation procedure, continued, if personal visit did not terminate the search. Further search then generally required professional survey evaluation and specific financial approval of the property before proceeding to the bid phase. Since most households were not shown the results of surveys in detail (and institutional control is probably excessive in the Scottish context) the reasons for

Table 6.9: Household Reasons for Not Commissioning a Survey Following a Personal Visit

| | |
|---|---|
| Disliked property | 40.6 |
| Poor area quality | 6.5 |
| Poor maintenance/decoration | 14.2 |
| No loan available | 6.3 |
| Wrong price | 14.1 |
| Coloured neighbours | 0.3 |
| Other | 4.8 |

an evaluation not resulting in a bid are somewhat vague. What is clear, is that some 34 percent of professional surveys did not generate a bid (see Maclennan and Jones, 1981). The number of surveys undertaken by eventually successful households reflects not only termination of a detailed search following an adverse survey but also mirrors the extent to which bids placed are not accepted.

Bids placed by the household reflect the searcher's resource constraints and judgement of the state of the market. But the households judgement regarding market conditions, since the price-offers by other searchers are unknown to him, is heavily influenced by solicitors or estate agents involved in placing bids. The way in which detailed survey information and price bids are concealed from potential purchasers must be a cause of concern in the Scottish housing system. The system is quite obviously informationally inefficient.

If the household's bid is not successful, then if the household still aspires to moving house, the search process has to be repeated at least from the post orientation phase. Table 6.10 shows the percentage distribution of repeat surveys. The mean value of bid/survey repetition exceeds 2.5 attempts per household. The direct costs of bidding for and surveying houses, aside from the unobserved costs on use of time, transport and channel use, are important (Figure 6.3). Ultimately repeated intensive searching may constrain the household's ability to finance house purchase.

## CONCLUSIONS

On the basis of the evidence presented in this essay, we would contend that an information acquisition and information use approach considerably enhances understanding of housing choice processes and outcomes. There are both specific and general advantages of this approach. In the case of the empirical examples presented here there are specific policy implications flowing from the insights gained from a search perspective. In the rental sector, with binding rent controls information and search became the critical rationing devices. For tenants many of the costs of rent controls are "hidden" in the search process and the control policy does not have a neutral effect insofar as there are information differences across groups. In the owner market in Scotland, the information/adjustment approach suggests that Adam Smith's "invisible hand" does not guide unknowingly, but rather gropes towards a solution in a context of restricted enlightenment. At stages of the search process the existing institutional framework allows lenders to express their area lending preferences without challenge from consumers and adherence to the sealed bid system protects the traditional housing market role of solicitors.

There are, of course, more general advantages of the market search process paradigm. Naturally, it emphasizes the problems of disequilibrium and adjustment which can be important in housing

Table 6.10: The Survey and Bidding Record for First-Time Purchase

| | Percentage of total sample wishing to instruct or survey on a house | Percentage of "intended" surveys actually undertaken | Percentage of "intended" surveys not undertaken on advice of institution |
|---|---|---|---|
| Present house | 100.0 | 87.1 | 6.9 |
| Previous house 1 | 48.0 | 51.0 | 47.5 |
| Previous house 2 | 25.0 | 59.0 | 39.9 |
| Previous house 3 | 13.9 | 37.5 | 54.0 |
| Previous house 4 | 7.2 | 54.0 | 46.0 |
| Previous house 5 | 3.8 | 46.2 | 53.4 |

| | Percentage of "intended" surveys resulting in a bid for property | Percentage of actual surveys not resulting in a bid | Proportion of bids made by a solicitor |
|---|---|---|---|
| Present house | 100.0 | 15.0 | 79.2 |
| Previous house 1 | 37.6 | 26.2 | 83.4 |
| Previous house 2 | 20.7 | 64.9 | 93.0 |
| Previous house 3 | 22.9 | 38.9 | 100.0 |
| Previous house 4 | 28.0 | 48.0 | 100.0 |
| Previous house 5 | 38.5 | 16.7 | 100.0 |

*Note:* The term "Previous house" refers to houses surveyed in the search sequence.

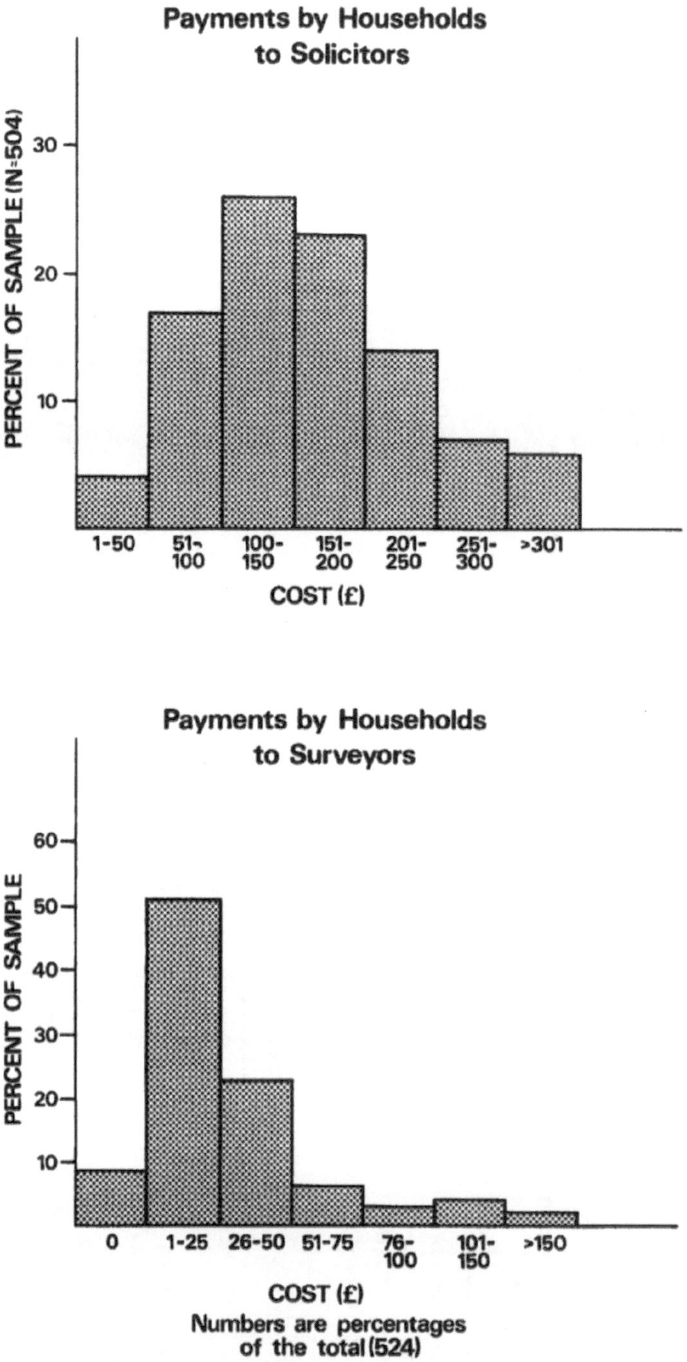


Figure 6.3. Payments by households to solicitors and surveyors.

markets. Housing market events or processes such as institutional "redlining" or racial discrimination may often be impossible to identify from equilibrium analysis of housing outcomes. But market process studies offer some hope of identifying such influences where they exist. Further, the costs of search now have to be explicitly considered as an influence on housing choices and expenditures. Studies of housing demand which omit such expenditures may generate inaccurate and biased estimates of price and income elasticities.

But there are also inherent research disadvantages of the search approach. The method of analysis requires detailed and expensive survey information. For many hypotheses to be tested ex ante as well as ex post information is required. This requires samples to be identified prior to or at the start of the search process. Then researchers have to rely on the accuracy of searcher memories and must avoid the pitfalls of ex post rationalisation. For many housing market purposes, where the market is thought to be in or near a position of equilibrium, or where policy is only to be crudely constructed, then this detailed and costly approach may be an inappropriate use of research resources.

If the information/search approach to housing choice is to be further developed then there are important research issues outstanding. Here we have examined individual behaviour but overall market clearing and the reaction and behaviour of information supply agents requires further investigation. At the level of the individual more refined modelling is required. For instance, it is appropriate to establish how aspiration levels and information network use are adjusted interactively and whether the decision stages or phases discussed above exist discretely or whether they merge and overlap. All these research issues are important, for as Kenneth Arrow has observed some time ago (Arrow, 1974) a reduction of uncertainty about the usefulness of the models of individual choice requires a more detailed and systematic understanding of how individuals react to uncertainty and imperfect information. This applied research agenda is still far from the point of diminishing returns.

## NOTES

1. This paper is based upon research data collected on projects supported by the University of Glasgow and the Scottish Office. The veiews expressed herein, are our own responsibility.

2. The available evidence on search processes does not conclusively support the notion of "search-stages". In this paper we do not attempt to precisely identify such stages. Rather, the stages identified here, arise from our a priori notion of the housing search process.

3. The costs of channel use are not precisely identified here. But they will include time and travel costs to searchers as well as direct payments for access to information.

4. A very serious modelling difficulty arises at this juncture. If the initial series of samples, using a single channel, generate vacancy, quality or price probabilities which suggest a prevailing distribution of opportunity different from those expected, two kinds of response are possible. The searcher may not alter their expectations about the housing market and, instead, revise expectations regarding message service quality. Alternatively, confidence in the message service implies a need to adjust housing market aspirations. In this paper we assume that initially households expand the range of message services before altering housing aspirations.

5. In neither of the empirical studies presented here is there any attempt to model information supply decisions by households. Clearly we would expect these patterns to vary with the state of the market. But both our market studies are characterised by sustained excess demand.

6. In the 1970's, in Scottish cities, local government had assumed considerable responsibility as a lender for house purchase in older inner city areas. (See Maclennan and Jones, 1981.)

## REFERENCES


Arrow, K. J. (1974) "Limited knowledge and economic analysis", American Economic Review, 64, 1-10.

Hirshleifer, J. and J. G. Riley (1979) "The analytics of uncertainty and information - an expository survey", Journal of Economic Literature, 1375-1421.

Johnson, J. H., J. Salt and P. A. Wood (1974) Housing and the Migration of Labor in England and Wales. London: Saxton House.

Lake, R. W. (1980) "Housing search experiences of black and white suburban homebuyers", in G. Sternlieb and J. W. Hughes (eds.) America's Housing: Problems and Prospects. New Brunswick, New Jersey: Rutgers University.

Maclennan, D. (1977) "Some thoughts on the nature and purpose of urban house price studies", Urban Studies, 14, 59-71.

----- (1978) "The 1974 Rent Act: some short-run supply effects", Economic Journal, 88, 331-340.

----- (1979 a) "Information and adjustment in a local housing market". Applied Economics, 11, 255-270.

----- (1979 b) "Information networks in a local housing market", Scottish Journal of Political Economy, 26, 73-88.

----- (1981 a) Housing Economics: An Applied Approach, Longmans, London: (Forthcoming).

----- (1981 b) "The 1974 Rent Act: some impacts on tenants". Social Policy and Administration, 15, 181-197.

----- and C. A. Jones (1981) "The Redlining Hypothesis Re-Examined", Glasgow University, (Mimeo).

Rees, A. E. (1966) "Information networks in labour markets", American Economic Review, 56, 559-566.
Reid, G. L. (1972) "Job search and the effectiveness of job finding methods", Industrial and Labour Relations Review, 9, 11-23.

# 7

# INFORMATION PROVISION: AN ANALYSIS OF NEWSPAPER REAL ESTATE ADVERTISEMENTS

**Terence R. Smith, W. A.V. Clark and Jun Onaka**

The information flow structures associated with housing markets are both critical and complex. They are critical insofar as they may strongly affect the decision-making behavior of all agents involved in housing market processes. They are complex insofar as they generally involve many channels through which diverse information with multiple characteristics may pass. Such channels include realty agents, newspaper advertisements, open houses, agents in financial institutions, and informal networks of friends and relatives. The characteristics of the information passing through the channels include content, representativeness, reliability, consistency, timeliness and cost.

It is possible to make a distinction between the supply and demand sides of information flow structures. Some care is required, however, in making this distinction. On the one hand, a given information flow channel may be on the supply side for some decision makers and on the demand side for other decision makers. On the other hand, a given decision maker may be on the supply side with respect to some channels of information flow and on the demand side with respect to other channels of information flow. Nevertheless, if used with care, the distinction between supply and demand sides is a useful one.

Much of the published research concerning housing market information flow systems has tended to emphasize the demand side. In particular, there has been a relative concentration of research on the manner in which prospective buyers obtain information from the various channels that provide information about vacancies in the market. There appears to be much less published research on the supply side, and the research available relates largely to the manner in which realty agents provide information.

It is maintained in this chapter that, while a great deal more research is warranted on housing market information flow systems in general, future research might profitably focus on the supply side of the information flow structure. In this chapter, therefore, we take a small step in the suggested direction and examine some of the characteristics of one particular channel that is important in the provision of information: the newspaper advertisement channel.

 DOI: 10.4324/9781003182085-9

The chapter is structured as follows. We first present a brief review of some of the relevant literature relating to the demand and supply sides of the housing market information flow system. We then discuss the reasons for examining the quantity of price and location information provided about housing vacancies in newspaper advertisements. Variations in the relative and absolute levels of price and location information are then related to variations in the goal structures and knowledge structure of advertisers at an individual decision-making level. To obtain a set of testable hypotheses, we differentiate between the behavior of owners and realty agents, and between advertisers in different market conditions. The advertisements that serve as the data base for testing these hypotheses are then described, together with the methodological basis for implementing the tests. Following this, we present the results of the hypothesis tests and draw some conclusions from the analysis.

## PREVIOUS RESEARCH

In this section of the essay, we briefly discuss some of the previous research relating to both the supply and demand sides of housing market information flow structures. It is to be emphasized that the section is not intended to be a complete review of past work, but only a brief description of recent research that has particular relevance for the present analysis.

### The Demand Side

Much of the research concerning the demand side of the information flow system has involved empirical investigations of the manner in which prospective buyers utilize various channels concerning housing vacancies. A tabular summary of much of this research, together with some general comments, is provided in Clark and Smith (1979) and some empirical research is reported in Clark and Smith (1982). Bettman et al. (1978) examined the demand for information in a sample of recent homebuyers. The results of their analysis indicated that there were several items of information that were perceived by buyers as having potential value, but that were generally perceived as being unavailable. These items related to neighborhood characteristics in general and to crime rates in particular.

There has been a relatively small amount of theoretical research relating to the demand side of the information system. This lack persists despite a substantial literature that examines rational decision-making models of information collection in the context of job market search. Clark and Smith (1979) and Smith and Clark (1980), however, have constructed and analyzed a decision-making model that describes how potential buyers decide upon the order in which various channels of information flow are to be consulted, and on the quantity of information to be sought from each channel. The channels include "newspapers", "friends and relatives", and "realty agents". The model is currently under investigation in an experimental setting, using subjects who have recently completed a housing market purchase. The purpose of the investigation is to seek the effects of different information channel cost structures on the patterns of channel usage.

The Supply Side

There has been a less pronounced research focus on the supply side of housing market information flows structures. Most of the published research has concerned empirical investigations of the provision of information by realty agents. Palm (1976a, 1976b) has examined the manner in which realty agents act as "filters" of the information provided to prospective buyers in parts of the San Francisco and Minneapolis housing markets. Two effects were investigated and found to be of significance. First, realty agents provided housing vacancy information in relation to the perceived personal characteristics of the buyers (such as professional status, income, etc.). Second, the information provided was biased in terms of the spatial location of the agent.

Houston and Sudman (1977) examined the quality and accuracy of information provided by realty agents to prospective buyers about neighborhood institutions. Houston and Sudman also examined the accuracy of information provided by realty agents concerning the characteristics of neighborhoods and the residents of the neighborhoods, using data obtained from a national sample survey. The quantity and accuracy of the information provided were assessed relative to the quantity and accuracy of information provided by "experts" from churches, schools, community organizations and information provided by long-established real estate brokers. In general, realty agents did not provide as high a quantity or quality of institutional information as the experts, but Houston and Sudman concluded that they did provide the best overall information concerning residential and neighborhood characteristics. Bettman et al. (1978) provide similar evidence concerning the quality and quantity of information provision.

A problem of interest raised by Houston and Sudman (1977) involves the question of legal constraints on the information provided by realty agents. A recent study by Palm (see Chapter 8) has examined this question in some depth in relation to the provision of earthquake hazard information. Such provision is required legally by the Alquist-Priolo Act in certain Special Study Zones located in the San Francisco Bay Area of California. Palm concluded that the Act is ineffective. First, it appears that conflict of interest on the part of realty agents may inhibit sufficient disclosure, although Palm views the realty agent as being the most appropriate channel for the provision of such information. Second, there appears to be little or no significant response in terms of either buyer behavior or housing prices to revealed earthquake hazard zones. Palm interprets such unresponsiveness as due in part to the relatively low salience of environment hazards, when compared with the salience of perceptions concerning inflation and financial investment.

Another source of information concerning housing vacancies that has current legal implications is the Multiple Listing Service (MLS) data. The importance of the MLS data as a source of information is reflected in at least one recent law case over access to MLS data. (Los Angeles Times, 1981) Whereas the initial cases addressing Boards of Realtors and multiple listing services were concerned with violations of antitrust legislation, especially with regard to uniform commission rates (People v. National Association of Realtors, 1981), more recent cases have raised the question of access to the MLS data. In one appellate court ruling, access to the residential portion of the MLS data was guaranteed to all licensed brokers without regard to membership. As the court noted, refusal of the right of access to any listing agreement other than one which grants the listing broker an exclusive right to sell, leads to an impermissible restraint on the public's ability to compete in negotiating for alternative type listings (People v. National Association of Realtors, 1981: 1931).

There are pending legal cases concerning the provision of MLS data to the public in general. The general provision of MLS data in edited form to the buying public could greatly alter the information supply structure of the housing market. Although MLS data does not represent all housing vacancies, it does provide an extensive coverage of most markets, and Becker (1972) has shown that there is no significant difference between MLS listings and a random sample of all vacancies in a given market.

As in the case of the demand side of the information flow structures, there has been little theoretical research concerning the supply side of the system. Smith and Mertz (1980) examined the manner in which the sequencing of vacancies shown by agents to prospective buyers could influence the particular vacancy chosen for purchase, and hence, could influence the price, characteristics, and ultimate "utility" of the vacancy to the buyer. Smith and Clark (1980), in considering their simplified model of a housing market information flow system, showed how certain criteria for efficiency of search could be maximized by the correct selection of a cost structure governing the usage of the various available information flow channels.

## The Need for Further Research

It is apparent that there are relatively few empirical or theoretical analyses of the demand side of housing market information flow structures and even fewer that consider the supply side of the structures. In particular, there is a need for studies that examine the information supplied to the various information flow channels.

# THE NEWSPAPER INFORMATION CHANNEL AND AMBIGUITY IN PRICE AND LOCATIONAL INFORMATION

In the present study we examine some of the characteristics of the information supplied by the newspaper advertisement channel. The determinants of these characteristics are examined in relation to simple models of individual decision making behavior.

Although the newspaper advertisement channel is not independent of other channels, it was chosen for a separate and preliminary analysis for three reasons. First, newspapers provide a source of information that is used early in the housing search process and that continues to be used throughout the process (Clark and Smith, 1979; Clark and Smith, 1982). Second, both the nature and content of the information provided by the channel are relatively easy to observe, although we have been unable to locate any research in which such an analysis has been performed. Finally, the information supplied by the channel has several characteristics of interest that include its representativeness of the market as a whole, the relatively low cost of its usage, and its relative timeliness. It is these characteristics that presumably give the newspaper advertisement channel salience during the early parts of a buyer's decision-making process.

## Quantity of Price and Location Information and Individual Decision Making

Newspaper advertisements may contain a great range of information concerning housing structural characteristics, location and neighborhood characteristics, financial arrangements and the selling agencies. In this analysis, we concentrate our attention on the price and location information provided in the advertisements. One reason for this concentration is that price and location are probably the two most salient characteristics of any vacancy, since both are dominant factors in housing evaluation, and hence, in purchase decisions (see Smith and Clark, 1982; Clark and Smith, 1982). Furthermore, a knowledge of both the price and location of a vacancy may provide a good deal of implicit information concerning housing quantity and quality, even if the vacancy is not viewed.

We also focus our attention on the quantity of information provided. This focus provides a second reason for concentrating on price and location information, since it is relatively easy to categorize the quantity of price and location information supplied in any advertisement. This ease of categorization does not extend to some of the other characteristics of the information provided, such as evaluative descriptors.

Our main goals in analyzing the quantity of price and location information supplied in newspaper advertisements are twofold. A first goal is to characterize the relative and absolute quantities of information provided about price and location in particular markets. A second, and more important, goal is to examine the determinants of any variations in the relative and absolute quantities of the information provided.

The main viewpoint that we adopt in explaining variations in the quantities of information provided is that the determinants of such variations may be viewed as arising at the individual decision-making level. The costs of advertising severely constrain the quantity of information that it is rational to provide in relation to a given vacancy. Hence individual advertisers are faced with a choice problem as to what information they should provide.

In order to implement this point of view we adopt a strategy of obtaining a set of simple, testable hypotheses from elementary considerations of the factors that enter into an advertiser's decision-making procedures. Observations on a set of newspaper advertisements are used as a basis for testing the hypotheses.

## Individual Decision making and the Ambiguity of the Price and Location Information Supplied

A first decision that an advertiser must make is whether a given vacancy is to be inspected after an appointment is set up between a prospective buyer and the owner (or agent) of the vacancy, or whether a vacancy is to be an open house. In the latter case, the advertisement contains an appointed time at which any prospective buyer may view the vacancy. The significance of this decision for the present analysis is that exact locational information (i.e., street name and house number) must (generally) be provided in the advertisement for an open house. This is not necessarily the case for non-open house advertisements. Since the outcome of the decision may constrain the provision of location information, a clear distinction must be made between open and non-open house advertisements in the present context.

A second decision that an advertiser must make concerns the relative quantities of price and locational information to provide. The outcome of this decision depends on whether the decision maker perceives price and locational information to be substitutes in some sense. According to many basic models of buyer decision-making behavior (e.g., Alonso, 1964), buyers are typically able to make tradeoffs between price and location in reaching a decision concerning which vacancy to purchase. It is not clear, however, that such "equivalence" exists between price and location information from the viewpoint of an advertiser. On the one hand, exact location information allows prospective buyers to view a vacancy without contacting the advertiser. Hence, in this case, exact location information may be interpreted as information concerning the nature of the vacancy. On the other hand, since exact price information effectively provides an upper bound on the bid prices expected for the vacancy, providing such information in an advertisement may be risky if the advertiser does not have a good knowledge of the distribution of bid prices.

In these terms, it is possible that a decision concerning the quantity of location information to provide represents in part a decision of how much ambiguity concerning housing characteristics to present to prospective buyers. The decision concerning how much price information to provide may represent how much ambiguity the advertiser has with respect to price distributions. It is not clear, therefore, to what extent price and location information are substitutes.

A third decision that an advertiser must make relates to the absolute levels of price and location information that are to be given. One may expect any differences between the levels of information provided by different individuals to depend on the goal structures and knowledge structures of the individuals. A potentially important difference in goal structures depends on whether the advertiser has only one vacancy to sell, or whether the advertiser has a set of vacancies to sell in addition to the advertised vacancy. In order to understand the significance of this difference for the purposes of the present analysis, it is necessary to consider the likely responses of a potential buyer who reads the advertisement. As the information provided in an advertisement is made more specific with respect to the characteristics of a given vacancy, it becomes easier for a reader to judge the likelihood that the vacancy would prove to be a suitable purchase, given the reader's preferences and beliefs. Since there are costs associated with inspecting a vacancy, this perceived likelihood will affect the likelihood that the individual will respond to the advertisement.

Given a market with both a large number of heterogeneous housing vacancies and a large number of prospective buyers with heterogeneous preferences and beliefs, one might expect reader response to ambiguity to vary as follows. On the one hand, a very high degree of specific information concerning the characteristics of the house might lead to a low response rate, with responding individuals having a relatively high probability of finding the vacancy acceptable. On the other hand, a very low degree of specific information concerning the vacancy might also lead to a low response rate, with responding individuals having a low probability of finding the vacancy acceptable. Intermediate degrees of ambiguity might lead to relatively high response rates, with respondents having an intermediate likelihood of finding the particular vacancy acceptable.

The preceding ideas may be related to the goals of selling either a single (advertised) vacancy or selling some vacancy from a portfolio of vacancies, only one of which is advertised. (It is assumed in both cases that there are significant costs involved in showing a vacancy to a respondent.) In the case of the advertiser attempting to sell a single vacancy, too much ambiguity about the characteristics of the vacancy might result in a relatively low likelihood of eliciting a response from a reader who will find the vacancy acceptable. Furthermore, the costs of showing the vacancy to respondents who find it unacceptable might be relatively high. In the case of the advertiser with a portfolio of vacancies, information that is too specific with respect to the advertised vacancy might cause readers who would find one of the vacancies in the portfolio acceptable to fail to respond.

One might expect, therefore, that the goals of the advertiser will determine the degree of ambiguity of the advertisement with respect to housing characteristics. If one now interprets location information in terms of information about the characteristics of the vacancy, one might expect variations in the level of location information to vary with the goal structures of the advertiser.

A potentially important difference in knowledge structures relates to the advertiser's perceptions of bid price distributions. In particular, if the individual feels sufficiently uncertain about bid price distributions, then the perceived risk of providing an unsuitable level of price information may be high. Advertised prices that are set too high might lead to few respondents. Prices set too low might result in a financial loss on selling the vacancy. Hence one might expect that the advertiser's degree of uncertainty concerning bid price distributions will be reflected in the level of price information given.

We assumed above that the determinants of variations in the quantity of information provided in advertisements may be viewed as arising at the level of an individual decision maker. At this level, however, environmental factors may affect the outcome of the advertiser's decision making. In particular, characteristics of the housing market such as size, heterogeneity, demand for housing, supply of housing and levels of housing market "activity" may affect the decision making process. Since these effects would be relatively uniform within a given market, we might expect inter-market variations to occur in the levels of ambiguity concerning price and location in the advertisements. It is reasonable to assume that in "active" markets, there might be relatively rapid changes in supply, in demand and in prices. These changes may result in bid price uncertainty on the part of advertisers, with a corresponding degree of price ambiguity written into the advertisements.

## HYPOTHESES CONCERNING THE QUANTITY OF PRICE AND LOCATION INFORMATION PROVIDED

We now present hypotheses concerning the determinants of the quantity of price and location information provided in advertisements. The hypotheses follow from the preceding arguments. Since these arguments are partly based on variations in the manner in which individuals make decisions, it is important to be able to categorize individuals with respect to their decision making activities. An extremely useful categorization involves the distinction between an owner selling a single vacancy and a professional realty agent with a portfolio of vacancies for sale. As noted above, we may expect the two classes of individuals to possess different goal and knowledge structures. Furthermore, it is possible to distinguish owners and realty agents when examining most newspaper advertisements. Hence we obtain one basis for constructing testable hypotheses.

A first hypothesis concerns the relative quantities of the price and location information given in advertisements. Although it is not clear on a priori grounds that advertisers should trade-off price and location information, and while it is possible to make arguments as to why price and location information should not be substitutable, we consider the hypothesis:

> H1: Both owners and realtors make substitutions between quantities of price and location information.

The next two hypotheses relate to individual differences, and both follow from the arguments of the preceding section.

> H2: Owners provide relatively more location information than realty agents.
>
> H3: Owners provide less price information than realty agents.

A final hypothesis concerns intermarket differences.

> H4: Both owners and realty agents increase the level of price ambiguity as the degree of uncertainty concerning bid price distributions increases.

In order to test the preceding hypotheses concerning the degree of ambiguity in price and locational information, and in order to explore advertising data for unhypothesized effects, the following overall research design was adopted. A set of newspaper advertisements from two cities were analyzed with respect to price and locational information for owners and for agents, in the cases of both open and non-open houses. Figure 7.1 represents the design. Four direct comparisons were examined, namely within-city variations between owners and agents for the two cities and between-city variations for both owners and agents.

It may be noted that there are significant phenomena for which we have no prior hypotheses. In particular, there are no hypotheses concerning the level of open house advertisements or the relationship of open house advertisements to the level of price information. Since it proved difficult to construct intuitively reasonable prior hypotheses concerning such phenomena, exploratory data analysis was employed in searching for unhypothesized effects.

## THE DATA AND TECHNIQUES FOR ANALYSIS

We now present some details concerning the data collected and the techniques employed for analyzing the data and testing the hypotheses.

| | Owner | Realtor |
|---|---|---|
| Stockton | EP<br>EL-OH<br>OH | EP<br>EL-OH<br>OH |
| Santa Barbara | EP<br>EL-OH<br>OH | EP<br>EL-OH<br>OH |

Figure 7.1. Design of the analysis. EP = exact price, EL-OH = exact location less open houses, OH = open houses.

## The Data

Housing advertisements were sampled over a 20-year period, 1960-1979, for Santa Barbara, Calaifornia, and over a 19-year period, 1960-1978, for Stockton, California. The criteria for choosing these cities included the facts that both are medium-sized cities with about the same population; that both possess only one major newspaper in which most advertisements appear; and that although both cities are in the same state, they appear to be characterized by different levels of housing market activity.

For both cities, all advertisements occurring on the Friday closest to the 15th May and the Friday closest to 15th November were examined in each year of the sampling. All price, locational, and owner/agent information was recorded for each advertisement and it was noted whether the advertisement referred to an open house. (When telephone numbers were given as the only information concerning owner or agent, a reverse telephone directory was used to determine whether the advertiser was an owner or an agent/broker.) Price information was classified into exact price or no price. The exact price class includes such descriptions as "low seventies," since sales price is typically negotiable. Location information was categorized into exact address or street address, general area, or no location. If realty agents were involved, the broker for whom the agents worked was recorded wherever possible. The reason for choosing Friday was to avoid the relatively small numbers of advertisements that characterize beginning days of the week, and the relative predominance of open house advertisements on weekends.

Other data characterizing the two housing markets was collected, as available, from local government agencies and from realty boards. Such data included population size and growth, the numbers of housing permits issued, the numbers of new dwellings built, the numbers of vacancies listed, the numbers of housing sales, and the value of housing sales.

Techniques of Analysis

Three dependent variable with binary values were used to characterize advertisements in the analysis reported here. The three pairs of binary values were exact price/non-exact price, exact location/non-exact location, and open house/non-open house. In order to test the hypotheses stated above, and in order to conduct an exploratory analysis of the data in cases where no prior hypotheses were established, two sets of statistical tests were employed. The first set of tests were designed to measure the degree of association between pairs of time series of dependent variables. The second set of tests were designed to characterize the nature of the arithmetic differences between pairs of series.

To examine the degree of association between pairs of series, 2 x 2 contingency tables for each period and for various pairs of dependent variables were considered with the third variable held constant. A correlation coefficient was estimated for each contingency table (see Bishop, Fienberg, and Holland, 1975). Hence, each pair of series was transformed into a single series of correlation coefficients. According to Bishop et al., these correlation coefficients have an asymptotically normal distribution, whose mean is equal to the population correlation coefficient.

The main strategy for examining the degree of association between series was to construct and examine confidence intervals for the estimated correlation coefficients, taking account of all the data in each series of correlation coefficients. Such confidence intervals were obtained by regressing the correlation coefficients against time in a quadratic regression of the form:

$$\rho_t = \rho + \alpha T_{it} + \beta T_{zt} + e_t \qquad (1)$$

where $\rho_t$ is the estimated correlation coefficient at time t, $\rho$ is an intercept term, $T_{it}$ is a measure linear in time, $T_{zt}$ is a measure quadratic in time, and $e_t$ is an error term. The theory of Bishop et al. (1975) assures us that $e_t$ possesses an asymptotically normal distribution. The linear and quadratic measures of time were constructed so that the intercept represented a good appoximation to the average value of the correlation coefficient.

In order to take into account heteroscedasticity and autocorrelation in the error terms $e_t$ of (1), two procedures were incorporated into the least squares estimating procedures for the parameters $\alpha$, $\beta$, and $\rho$. First, each observation on equation (1) was modified by multiplying the dependent and independent variables by a deflator, related to the number of advertisements examined at each sample point. These numbers varied between samples. Deflators were constructed using the expression for the variance of the sampling distribution of the estimates of the correlation coefficients given by Bishop et al. (1975). Such deflators remove one correctable source of heteroscedasticity. Second, a Cochran-Orcutt iterative procedure was employed in the estimation of $\alpha$, $\beta$, and $\rho$ in order to take account of any first order autoregressive effects in the disturbance terms $e_t$ (see Maddala, 1977).

In interpreting the results of the regressions (1), the absence of statistically significant estimates of $\alpha$ and $\beta$ was taken to imply a constant value of the correlation coefficients relating any pair of time series. Hence, it was sufficient to consider whether the value of $\rho$ was significantly different from zero in order to test for association between series. With significant trend terms ($\alpha$ and $\beta$ significantly different from zero), it became necessary to examine confidence intervals about the regression line in order to test for periods of observation during which the measures of association were significantly different from zero.

In order to test for significant differences between any pair of comparable series, the arithmetic differences of the series were computed and Cochran-Orcutt estimation procedures were applied to equations of the form:

$$\delta_t = \delta + \alpha T_{it} + \beta T_{zt} + e_t \qquad (2)$$

where $\delta_t$ is the difference between the series at time t, $\delta$ represents a good approximation to the average value of the difference, and $\alpha$, $\beta$, $T_{it}$, $T_{zt}$, and $e_t$ have the same interpretation as in equation (1).

The absence of significant values of $\alpha$ and $\beta$ was taken to imply a constant value of the differences between any pair of series. Hence, a hypothesis of significant differences could be tested by examining whether $\delta$ was significantly different from zero. In the case of significant trend terms $\alpha$ and $\beta$, it was necessary to examine confidence intervals about the predicted values of the series of differences in order to determine periods during which the differences were significantly different from zero.

For all tests of both association and difference between series, values of the Durbin-Watson statistic indicated that the assumption of first order autoregressive disturbances was sufficient to take account of autoregressive effects in the disturbances.

## RESULTS OF THE ANALYSIS

In this section, we characterize the differences between the Stockton and Santa Barbara housing markets, and present data characterizing the advertisements that were analyzed. We then present the results of tests of the four hypotheses, as well as the results of an exploratory analysis in which unhypothesized effects were sought.

### Market Conditions in Stockton and Santa Barbara

Tables 7.1 - 7.3 contain data characterizing some of the demographic and housing characteristics for the two cities, summarized by five-year periods. Table 7.1 shows population in terms of absolute numbers and percentage changes. While both cities are approximately of the same size, the patterns of growth have been quite different. Santa Barbara exhibited a significant decrease in the rate of growth after 1970. In Table 7.2, we present data on the construction of new, single family homes. Stockton exhibits a generally accelerating accretion to its housing stock. It should be noted that Santa Barbara, however, was characterized by severe growth restrictions after 1970, with building moratoria in the areas of Goleta and Montecito, and with downzoning in the City of Santa Barbara.

Table 7.3 provides some information concerning the housing market activity. The average number of vacancies listed per month in the Multiple Listing Service (MLS) of each city have increased in each five-year period, with the numbers of multiple listings in Santa Barbara exceeding those in Stockton. On the other hand, the numbers of sales per month in Santa Barbara exceeded the number of sales in Stockton only after 1970. The average price per sale was higher for each of the four periods in Santa Barbara, and especially in the last five-year period.

If used with care, one may interpret the number of multiple listings per sale as a rough measure of average time on the market for vacancies. Hence, this ratio and the number of listings per household provide measures of market "activity." Assuming that Stockton and Santa Barbara are roughly equivalent in size, we note from Table 7.3 that in the first ten years of study, 1960 - 1969, Stockton had fewer listings per sale and fewer listings than Santa Barbara, while in the second ten-year period, Stockton had more listings per sale and fewer listings than Santa Barbara. It should be noted that the second period was characterized by strong building controls in Santa Barbara. Hence, we may tentatively conclude that the Stockton market was at first more active and then less active than the Santa Barbara market.

Table 7.1: Population of Stockton and Santa Barbara

| | Stockton | | | | Santa Barbara | | | |
|---|---|---|---|---|---|---|---|---|
| | City | | SMSA | | City[a] | | SMSA | |
| | Pop. | Ave. Ann. % Change | Pop. | Ave. Ann. % Change | Pop. | Ave. Ann. % Change | Pop. | Ave. Ann. % Change |
| 1960 | 86.3 | --- | 155.6 | --- | 77.8 | --- | 169.0 | --- |
| 1965 | 97.1 | 2.5 | --- | --- | --- | --- | --- | --- |
| 1970 | 110.0 | 2.7 | 176.2 | 1.3 | 130.4 | 6.8 | 264.3 | 5.6 |
| 1975 | 118.0 | 1.5 | --- | --- | 141.9 | 1.8 | --- | --- |
| 1980 | 149.8 | 5.4 | 205.0 | 1.6 | 144.2 | 0.3 | 298.7 | 1.3 |

*Source:* U.S. Census of Population and City Planning Departments of Stockton and Santa Barbara.

*Note:* Population figures are in thousands. Average annual rate of change applies to previous 5 or 10 years according to data availability.

[a]includes incorporated Santa Barbara and Goleta.

Table 7.2: Average Number and Value of New Single Family Homes Constructed Per Year in Stockton and Santa Barbara by Five-Year Periods.[a]

| | Stockton | | Santa Barbara[b] | |
|---|---|---|---|---|
| | No. of Units | Value/ Unit | No. of Units | Value/ Unit |
| 1960 - 64 | 539.0[c] | 10.6[c] | 2287.6[e] | 18.0 |
| 1965 - 69 | 408.3[d] | 20.6 | 804.0 | 25.77 |
| 1970 - 74 | 657.4 | 24.5 | 725.4 | 34.8 |
| 1975 - 79 | 1384.8 | 46.8 | 844.2 | 64.7 |
| Average All Years | 598.8 | 36.4 | 1041.2 | 33.5 |

*Source:* California Construction Trends, Los Angeles, California: Security Pacific Bank.

*Notes:* [a]Value per unit in thousands of dollars.

[b]Includes unincorporated Santa Barbara County.

[c]Based on 1960 only.

[d]Based on 1967 - 69 only.

[e]Based on 1962 - 64 only.

## The Numbers of Advertisements and the Proportions with Different Price and Locational Information

Although the major concern of the analysis lies with the proportion of advertisements supplying different types of information, we first present some data characterizing the absolute numbers of advertisements, averaged over five-year periods, in Table 7.4. First, it may be observed that for the whole period, about one advertisement in every six was placed by an owner in Stockton, while the equivalent ratio for Santa Barbara is about one in seven. The temporal pattern in the numbers of advertisements placed by realty agents is the same for both cities, although Stockton shows greater percentage declines in the first three of the five-year periods and

Table 7.3: Average Number of Multiple Listings Per Month, Sales Per Month, Listings Per Sale, and Price Per Sale (thousands of dollars), by City and Five-Year Periods.

| | Stockton | | | | Santa Barbara | | | |
|---|---|---|---|---|---|---|---|---|
| | Listings | Sales | Listings/ Sale | Price/ Sale | Listings | Sale | Listings/ Sale | Price/ Sale |
| 1960 - 64 | 132.8[a] | 84.0[a] | 1.58[a] | 15.2[a] | 184.0[a] | 83.0[a] | 2.22[a] | 21.7[a] |
| 1965 - 69 | 202.6 | 108.4 | 1.87 | 15.9 | 261.0 | 99.2 | 2.63 | 27.0 |
| 1970 - 74 | 241.8 | 99.8 | 2.42 | 26.3 | 341.2 | 209.6 | 1.63 | 35.2 |
| 1975 - 79 | 380.5[b] | 222.5[b] | 1.71[b] | 38.3[c] | 441.0[d] | 255.0[d] | (1.59)[e] | (91.4)[e] |
| Average All Years | 225.0 | 119.3 | 1.89 | 25.5 | 209.4 | 106.6 | 1.97 | ---[f] |

*Source:* California Real Estate (Multiple Listings Reports), Los Angeles, California: California Association of Realtors. Data are for April of each year, except where noted.

*Notes:* [a]Based on 1960 - 63 only.

[b]Based on 1975 - 78 only.

[c]Based on 1975 - 77 only.

[d]Based on 1975 only.

[e]Based on annual data for 1975 - 79.

[f]Not computed due to lack of comparable data.

Table 7.4: Average Number of Advertisements Per Day by City, Agent, and Five-Year Periods.

| | Stockton | | Santa Barbara | |
|---|---|---|---|---|
| | Owners | Realtors | Owners | Realtors |
| 1960 - 64 | 31.8 | 143.7 | 23.6 | 158.9 |
| 1965 - 69 | 22.7 | 120.9 | 31.3 | 143.3 |
| 1970 - 74 | 16.9 | 84.8 | 26.6 | 139.5 |
| 1975 - 79 | 22.4[a] | 89.9[a] | 27.4 | 197.1 |
| Average All Years | 23.5 | 110.9 | 27.2 | 159.7 |

*Note:* Entries are averages of the number of advertisements in two editions of daily newspapers, one in May and one in November, for the years indicated.

[a]Data for 1975 - 78 only.

Santa Barbara shows a much larger percentage increase in the last of the five-year periods. Santa Barbara also has twice the number of listings occurring in Stockton during the last of the five-year periods.

In Figures 7.2 and 7.3, we present time series data that characterize the newspaper advertisements in terms of the proportions of advertisements giving exact price and location information. The four subfigures represent the four cases of owner and realty agents by Stockton and Santa Barbara. In each subfigure, the proportions of advertisements giving exact price, exact location, and open house are shown for each year, averaged over the spring and fall observations.

There are several general comments that can be made about these series. First, we found no significant seasonal differences between the May and November observations for either town. In particular, the time series do not display the significant one-lag negative auto-correlations that one would expect of bi-seasonal data. A possible reason for this lack of seasonality is that the data are represented in terms of proportions. (The absolute numbers of advertisements, on the other hand, do display seasonal effects.) We assume the absence of seasonal effects in these proportions for the

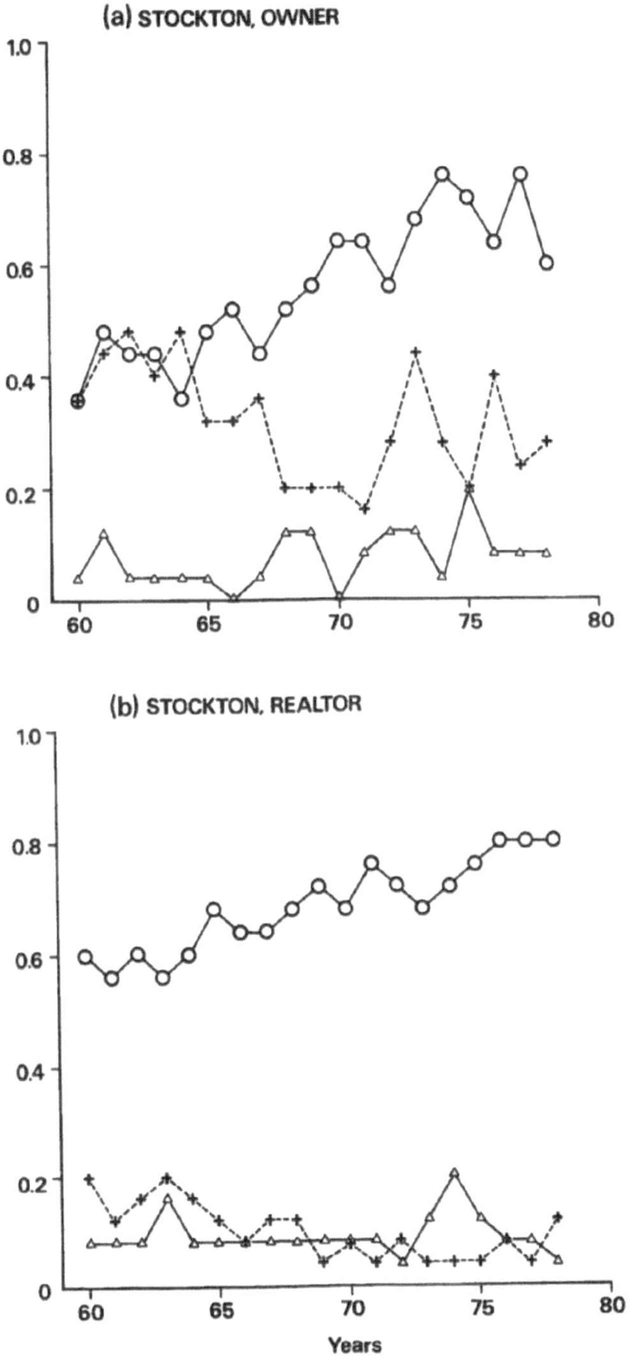


Figure 7.2. Time series plots of information types (proportions), 0 = exact price, + = exact location less open houses, Δ = open houses.

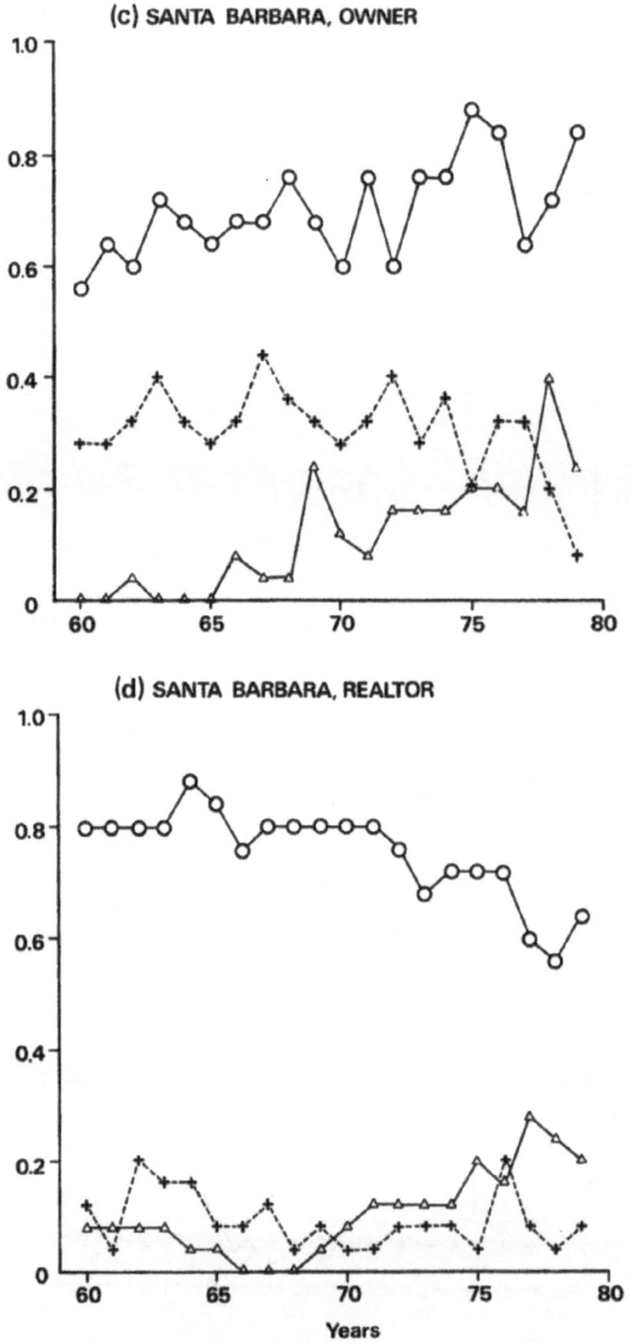


Figure 7.3. Time series plots of information types (proportions), O = exact price, + = exact location less open houses, Δ = open houses.

remainder of the analysis. Second, in all four cases, exact price information is given more frequently than exact location or open house information, and usually occurs in more than 50% of the advertisements. Exact location is usually given more frequently than open houses, which generally make up less than 20% of the advertisements. (It should be recalled that this data only characterizes the advertisements given on Fridays and that there are usually more open houses advertized on weekend days.)

## The Outcome of Testing the Hypotheses

The first hypothesis H1 states that advertisers can achieve a given degree of ambiguity by trading-off price and location information. The main results of an analysis of association between price and location information are presented in Table 7.5. For the case of realty agents in Stockton, the correlation coefficients are stationary, while the mean correlation coefficient is negative and significantly different from zero. For the case of realty agents in Santa Barbara, there is a significant upward trend in the correlation coefficients over time. The average correlation coefficient is negative and significantly different from zero. In the last few years of the 20 year period, however, individual correlation coefficients became significantly positive. For owners in Stockton, the average correlation coefficient is positive and significantly different from zero, although individual correlation coefficients are negative at the beginning and the end of the period. For owners in Santa Barbara, the series of correlation coefficients shows no significant trend over time, and the average correlation coefficient is not significantly different from zero.

In summary, the results for realty agents suggest that significant trade-offs are made between price and location information in achieving a desired level of ambiguity, although the positive relation between price and location information towards the end of the period in Santa Barbara confuses the issue. Analogous results for owners indicate that these decision-makers do not make such trade-offs. For Stockton, in fact, owners who give exact price information tend to give exact location information. Hence, the hypothesis H1 is essentially confirmed for realty agents but not for owners.

The second hypothesis H2 states that owners provide more location information than realty agents. An analysis of the differences between proportions of advertisers giving exact location information indicates that for Stockton, there is a significant quadratic trend in the differences (Table 7.6). On average, owners provide significantly more location information than realty agents. The same results apply in Santa Barbara. Hence, the hypothesis H2 is confirmed.

The third hypothesis H3 states that owners provide less price information than realty agents. For Stockton, there is no significant trend in the series of differences relating to exact price information.

Table 7.5: Tests of Association Between Series

| Association Between | Town | Advertiser | No. of Obs. | t-statistic average correlation coefficient | t-statistic linear trend coefficient | t-statistic quadratic trend coefficient |
|---|---|---|---|---|---|---|
| EP[a], EL-OH[b] | Stockton | Owner | 37 | 2.15 | 2.15 | -2.42 |
| EP, EL-OH | S. Barbara | Owner | 35 | -0.35 | 1.21 | -1.49 |
| EP, EL-OH | Stockton | Realtor | 37 | -2.73 | -0.46 | 0.61 |
| EP, EL-OH | S. Barbara | Realtor | 39 | -5.34 | -1.73 | 2.82 |
| EP, OH[c] | Stockton | Owner | 15 | -1.33 | -2.23 | 2.35 |
| EP, OH | S. Barbara | Owner | 25 | -0.62 | -0.29 | 0.52 |
| EP, OH | Stockton | Realtor | 37 | -4.78 | 1.15 | -1.48 |
| EP, OH | S. Barbara | Realtor | 36 | -1.89 | 3.10 | -3.09 |

*Notes:* [a]Proportion of all advertisements by agent giving exact asking price (EP).

[b]Proportion of all advertisements by agent giving exact location, but excluding open houses (EL-OH).

[c]Proportion of all advertisements by agent with open house (OH).

Table 7.6: Tests of Mean Differences Between Series

| Difference Between | City/ Advertiser | Dependent Variable | No. of Obs. | t-statistic for average difference | t-statistic linear trend coefficient | t-statistic quadratic trend coefficient |
|---|---|---|---|---|---|---|
| Owner/ Realtor | Stockton | EP[a] | 37 | -5.71 | 0.58 | -0.27 |
| | | EL-OH[b] | 37 | 12.40 | -2.20 | 1.92 |
| | | OH[c] | 37 | -1.47 | -0.10 | 0.39 |
| | S. Barbara | EP | 39 | -4.20 | -0.52 | 2.51 |
| | | EL-OH | 39 | 15.99 | 3.38 | -3.83 |
| | | OH | 39 | 1.11 | 3.03 | -2.57 |
| Stockton/ S. Barbara | Owner | EP | 37 | -5.37 | 0.77 | -0.08 |
| | | EL-OH | 37 | -0.34 | -3.51 | 3.21 |
| | | OH | 37 | -2.91 | -0.94 | -0.23 |
| | Realtor | EP | 37 | -7.71 | -0.21 | 3.70 |
| | | EL-OH | 37 | 0.03 | -0.23 | -0.10 |
| | | OH | 37 | -0.61 | 2.75 | -4.14 |

*Notes:* [a]Proportion of all advertisements by agent giving exact asking price (EP).

[b]Proportion of all advertisements by agent giving exact location, but excluding open houses (EL-OH).

[c]Proportion of all advertisements by agent with open house (OH).

Information provision

Owners provide significantly less price information than realty agents. For Santa Barbara, there is a significant positive trend in the series. The average difference over the whole period is significantly negative, indicating that owners provide less information than realty agents. Towards the end of the 20 year period, however, owners provide significantly more price information than realtors. Hence, the hypothesis is only partially supported by the Santa Barbara data.

The fourth hypothesis H4 states that both owners and realty agents increase the level of price ambiguity as uncertainty concerning bid price distributions increases. We argued above that Stockton may have been the more active market in the 1960's, and Santa Barbara the more active market in the 1970's. The differences between the Stockton and Santa Barbara series for proportions of exact price advertisements placed by realty agents shows a significant positive trend over time, with Santa Barbara realtors giving significantly more price information in the 1960's, and significantly less price information in the 1970's. These results tend to support the hypothesis. For owners, there is a positive but insignificant trend in the series of differences. Owners in Stockton give significantly less price information in the early period, while towards the end of the period, the differences become insignificant. Again, the data provide support for the hypothesis.

In conclusion concerning the hypotheses, H1 is supported for realtors, but not for owners; H2 is supported for both Stockton and Santa Barbara; H3 is supported for Stockton, but only partially for Santa Barbara; and H4 is supported for both realtors and owners. While some of the trends in both the relations between series and the differences between series are difficult to interpret, the regularity in both sets of series is of great interest, indicating the possible existence of "regular" causative factors underlying the observed phenomena.

The Outcome of Exploratory Data Analyses

For several phenomena of interest, it was not possible to deduce reasonable hypotheses, particularly with respect to open house advertisements. We now present the results of an exploratory analysis of some of the time series of correlation coefficients and differences that were not examined in testing the four hypotheses considered previously.

We first consider the association between exact price information and open houses. Only advertisements providing exact location information were considered. The series of correlation coefficients characterizing the advertisements placed by realty agents in Stockton shows no significant trend. Exact price information is inversely and significantly related to open houses. In other words, realtors give less price information when advertising an open house than when advertising a vacancy for which exact location is given. The same result holds for Santa Barbara although there is a significant quadratic trend in the series of correlation coefficients.

For owners, there is no significant association in either the Stockton or Santa Barbara data.

Turning now to a consideration of the proportions of open house advertisements, the data for Stockton indicate that there is no significant difference between the proportion of open houses advertised by realty agents and by owners. (See Table 7.7 for summaries by five year periods). Furthermore, the series of differences shows no significant trend. In Santa Barbara, the average difference between the same series is also insignificant, but the series show a quadratic trend, with realtors advertising a higher proportion of open houses than owners in the early part of the period.

Finally, we consider differences between Stockton and Santa Barbara in relation to the advertising of open houses. While the average difference in proportions of open houses advertised by realtors is not significant between the two cities, the difference decreases significantly over time, and becomes significantly negative in the later period. Hence, Santa Barbara realty agents advertise more open houses than Stockton realtors. While the trend is also negative for owners, the effect is not significant. However, Santa Barbara owners advertise a significantly higher proportion of open houses than Stockton owners in the latter part of the period, and a significantly greater proportion on average for the whole period (Table 7.7).

If we interpret these results in the light of hypothesis H4, the negative relation between exact price information and open house advertisements for realtors in both Stockton and Santa Barbara indicates that open houses may be employed to a greater extent in periods of greater price uncertainty. This idea is strengthened by the fact that in the latter part of the period of study, open house advertisements by realtors increase significantly in Santa Barbara compared to Stockton (Table 7.7). This latter period was hypothesized to be a period fo relatively more price uncertainty in Santa Barbara than in Stockton. This effect is weakly shadowed by the data characterizing owners. It should be noted, however, that the expected difference betwen realtors and owners does not materialize, since one would expect owners to advertise more open houses than realtors if price uncertainty were the only factor. On the other hand, the costs of open houses for owners may be relatively high.

## CONCLUSIONS

The analysis of the quantity of price and location information provided about housing vacancies in newspaper advertisements has been shown to provide a rich basis for explaining several aspects of the supply side of the housing market information system. The significant differences in behavior between realtors advertising one vacancy from a portfolio of vacancies and owners advertising a single vacancy may be interpreted as affects arising from the variation in both individual goal structures and individual knowledge structures.

Table 7.7: Proportion of Advertisements Placed by Owners and Realtors Which Contain Various Information Types, by City and Five-Year Periods.

| | Stockton | | | Santa Barbara | | |
|---|---|---|---|---|---|---|
| | Exact Price | Exact Location Not Open House | Open House | Exact Price | Exact Location Not Open House | Open House |
| **Owners** | | | | | | |
| 1960 - 64 | 42.5 | 41.8 | 5.7 | 65.3 | 31.4 | 0.4 |
| 1965 - 69 | 48.5 | 28.6 | 5.7 | 67.1 | 34.2 | 7.7 |
| 1970 - 74 | 63.3 | 24.3 | 7.7 | 71.1 | 32.7 | 12.8 |
| 1975 - 79 | 68.8 | 27.2 | 10.3 | 75.2 | 23.4 | 25.9 |
| Average All Years | 53.2 | 32.3 | 6.8 | 69.7 | 30.5 | 11.9 |
| **Realtors** | | | | | | |
| 1960 - 64 | 58.7 | 16.6 | 9.5 | 82.3 | 14.5 | 6.6 |
| 1965 - 69 | 66.9 | 10.2 | 7.6 | 79.4 | 8.4 | 2.6 |
| 1970 - 74 | 71.2 | 6.1 | 10.6 | 74.9 | 6.5 | 10.5 |
| 1975 - 79 | 80.2 | 7.6 | 7.6 | 64.4 | 9.3 | 21.7 |
| Average All Years | 67.3 | 11.1 | 8.8 | 74.5 | 9.8 | 11.2 |

*Note:* Entries are percentages.

Owners provide significantly more location information and significantly less price information than realty agents. These effects may be related to the owner's goal of selling only a single vacancy, and to the owners relative lack of knowledge concerning bid price distributions.

The differences between the quantity of price information provided in the newspapers of two towns may be ascribed to differences in market conditions. In particular, bid price uncertainty may lead to more ambiguity in the price information provided. The evidence examined suggests that the proportion of advertisements relating to open houses may also be related to bid price uncertainty in the market. In particular, open houses and bid price uncertainty may be directly related.

The available evidence also suggests that realty agents may treat price and location information as (partial) substitutes to optimize the degree of ambiguity that will tempt readers to respond to advertisements. The same substitutability does not appear to hold for owners. This suggests that exact location information may be related to the provision of detailed information on housing characteristics as far as owners are concerned, while price ambiguity is related to an owner's relative lack of knowledge concerning bid price distributions.

We conclude by suggesting that the remarkable regularities in both the cross-sectional and time series data analyzed in this essay indicate that further, more detailed examinations of newspaper housing advertisements may prove of great value. In particular, a more detailed investigation of the individual decision-making processes that underlie the published advertisements appears to be highly warranted. Furthermore similar analyses of other channels on the supply side of the housing market information flows system would prove of great value.

## REFERENCES


Alonso, W. (1964) Location and Land Use. Cambridge, Mass.: Harvard University Press.

Becker, B. W. (1972) "On the reliability of multiple listing service data", The Appraisal Journal", 40, 264-267.

Bettman, J. R., N. Capon, R. J. Lutz, G. E. Belch and M. Burke (1978) "Affirmative disclosure in home purchasing", Los Angeles, California: University of California Housing, Real Estate, and Urban Land Studies Program, Occasional Paper 14.

Bishop, Y. M., S. E. Fienberg and P. W. Holland (1975) Discrete Multi-Variate Analysis, Cambridge, Mass.: MIT Press.

Box, G. E. P. and G. M. Jenkins (1970) Time Series Analysis. San Francisco: Holden Day.

Clark, W. A. V. and T. R. Smith (1979) "Modelling information use in a spatial context", Annals of the Association of American Geographers, 69, 575-588.

----- and T. R. Smith (1982) "Housing market search behavior, and expected utility theory II: The Process of Search", Environment and Planning A, (forthcoming).

Houston, J. M. and S. Sudman (1977) "Real estate agents as a source of information for home buyers", The Journal of Consumer Affairs, 11, 111-121.

Los Angeles Times (1981) 17th July.

Maddala, G. S. (1977) Econometrics. New York: McGraw Hill.

Palm, R. (1976a) "Real estate agents and geographical information", Geographical Review, 66, 267-280.

---- (1976b) "The role of real estate agents as information mediators in two American cities", Geografiska Annaler, 5B, 28-41.

People v. National Association of Realtors et al. (1981) Daily Journal D. A. R. (1929) (C. A. 4th, 16 June 1981).

Smith, T. R. and W. A. V. Clark (1980) "Housing market search: information constraints and efficiency", in W. A. V. Clark and E. G. Moore (eds.), Residential Mobility and Public Policy, Beverly Hills, CA: Sage Publications.

----- and R. Mertz (1980) "An analysis of the effects of information revision on the outcome of housing market search, with special reference to the influence of realty agents", Environment and Planning A, 12, 155-174.

----- and W. A. V. Clark (1982) "Housing market search behavior and expected utility theory I: measuring preferences for housing", Environment and Planning A, (forthcoming).

# 8

# HOMEBUYER RESPONSE TO INFORMATION CONTENT

**Risa Palm**

As the interest in the ways in which households make residential relocation decisions has increased, there has been a concomitant interest in the role of information as it affects those decisions. Even though the role of information in household decision-making is still only vaguely understood, there are three broad approaches to achieving a better grasp of how different amounts and kinds of information can affect the household relocation decisions. First, there have been several descriptive analyses of information sources (see Clark and Smith, 1979, for a review). Second, there have been a more limited number of studies which have attempted to model information source usage (Smith and Clark, 1980; McCarthy, 1979). Third, there have been studies of specific information usage by real estate agencies -- the major actors in the relocation decision process (Palm, 1976). This chapter is concerned with an analysis of the interconnected effect of federal intervention in information provision by the real estate agent.

In recent years, there has been increased federal and state regulation of house sales transactions. Some of this regulation is aimed at ensuring that prospective buyers are more fully aware of the total costs and risks associated with the home purchase process. For example, federal regulations now require complete disclosure of mortage loan costs, including detailed estimates of closing or transaction costs; and state court decisions have mandated full disclosure of all material facts by real estate agents before the consummation of a purchase contract. Similarly, environmental information, that concerning hazard potential emanating from proximity to a floodplain, unstable slope conditions which might result in landsliding under certain conditions, or proximity to active fault traces which might cause surface rupture in the event of even a minor earthquake, has also been provided to homebuyers under a variety of federal and state regulations. The federal government

*AUTHOR'S NOTE: This material is based upon work supported by the National Science Foundation under grant PFR78-04775. Any opinions, findings, and conclusions or recommendations expressed herein are those of the author and do not necessarily reflect the views of the National Science Foundation.*

DOI: 10.4324/9781003182085-10

requires that lenders notify prospective borrowers that property is located in a flood hazard area as defined by the Federal Insurance Administrator, when communities are part of the federally subsidized flood insurance program.

An even more complete disclosure requirement is in force in Santa Clara County, California, where the board of supervisors requires sellers of property partly or wholly within flood, landslide and fault-rupture zones to provide a written statement of geologic risk to prospective homebuyers. Similar, though less sweeping legislation (the Alquist-Priolo Special Studies Zones Act) requires that the real estate agent (or the seller, if not represented by an agent) to disclose to a prospective buyer the fact that the property is located within the special studies zone (a fault-rupture zone) as defined by the state geologist. Such legislation has been designed to increase the information available to prospective homebuyers, on the assumption that individuals have the right to know the risks they are assuming in buying a particular house in a specific location. In addition, it is presumed that with provision of such informaiton, buyers will be better able to make a decision concerning the physical hazards of the site, under conditions of caveat emptor. Although such legislation appears to fill a needed gap in consumer protection and was assumed to be working to convey "complex hydrologic, seismic, and other geologic information...to real-estate buyers before the sale" (Kockelman, 1980:71), it was not known whether such information actually affected the behavior of homebuyers. Does information disclosure alone (of environmental hazards in this case) have an impact on the decision-making process of homebuyers? Is information disclosure, as presently mandated, a sufficient method of influencing buyer decisions or subsequent mitigation measures? The essential questions being raised in this analysis are the extent to which consumer (homebuyer) responses are being modified by the disclosure of information, and the extent to which consumers (homebuyers) are being protected by legislation which requires that real estate agents disclose environmental hazards information.

## CONDITIONS UNDER WHICH BUYER RESPONSE COULD BE EXPECTED

Before analyzing public response to mandated environmental hazards disclosure, it is critical to specify the conditions under which one could reasonably expect a specific buyer response. A "measurable response" in this case can be defined as an attempt by the buyer (1) to avoid or modify the purchase contract, (2) to choose not to move to the area because of the negative information received, (3) to attempt to bargain with the seller for a lower price or better contract terms, or (4) to purchase the property at the original contract price, but to take measures to mitigate damage from any subsequent earthquake. Such mitigation responses might include purchasing earthquake insurance, storing food, water, or other supplies, making structural changes and reinforcements to the dwelling unit, or planning for family and community responses in the event of an earthquake. In other words, the prospective buyer may respond to the disclosure of environmental hazards by avoiding the house purchase or modifying the purchase terms (trading off environmental risks for economic benefits), or by preparing himself and his family against financial losses, property damage, and personal injury. Any of these responses indicate that the buyer has learned more about the environment, and modified his behavior accordingly.

The ways in which there might be a measurable response to negative environmental information can be classified by characteristics of the information receiver (the buyer), characteristics of the message (accuracy and consistency), characteristics of the senders (agents) and the context in which the message occurs. Table 8.1 summarizes the possible influences in each of these categories and they are elaborated briefly in the following paragraphs.

First, a receiver (buyer) must be open or motivated to respond to the information transmitted. If environmental hazards are of importance to a buyer in the overall purchase decision, then negative environmental information should have an influence on his initial purchase decision and subsequent mitigation decisions. If, on the other hand, environmental hazards have a low priority to homebuyers, then disclosure of such hazards will probably have little impact.

Another reason why disclosure may have little impact on the buyer, even if environmental factors are of particular importance in the buyer's purchase decision and the data are clearly presented, is that the buyer may not feel he has the latitude to take alternative action. Before the buyer can respond to hazards information (1) there must be alternatives available to the buyer other than simply purchasing the housing and absorbing the associated risk; (2) the buyer must be aware of these alternatives; (3) the buyer must assess the alternatives as feasible and actually advantageous; and (4) he must have the opportunity to adopt these measures. These conditions can be complex and difficult to assess objectively, but can be exemplified in the extreme case by the individual who has only $5,000 to invest as a downpayment and can afford only $450 a month for monthly payments; has few houses to choose from, all of which are in the hazards zone; has been told that no hazard insurance is available at a reasonable cost, or has been told that hazard insurance is not economically feasible because of high premium costs and high deductibles; and must make a decision among available alternatives within a ten-day period. In such a case, the prospective buyer would have little alternative but to purchase a house in the hazards zone, and forego hazards insurance. Although he might be worried by the hazards disclosure, there would be little he could do to modify his behavior in light of this information. Finally, a number of individual and cultural factors may influence the response of the homebuyer to disclosure. He may be particularly risk-prone, and willing to tolerate a high probability of property damage. Or he may be fatalistic, feeling that there is little he can do to avoid or mitigate damage from environmental hazards.

Second, even if environmental information is considered important in the location decision, disclosure may still have little impact on the buyer. This can occur if the information which reaches the buyer is confused, biased or in other ways unclear. For example, the message may be contradicted with other persuasive information, or incorrectly reported. As an example of the latter, the basic information that a house was located within a "one hundred year floodplain" (meaning that every year there is a one percent chance that the property would be flooded) could be presented as a message in which a flood would only occur once in a hundred years which will lead to an inaccurate or unclear perception of risk and the probability of buyer response will be accordingly reduced. Similarly, a written disclosure can be mediated by oral contradictions ("Sign this, but I want you to know that this area has never had a flood in the thirty years I've lived here").

Third, the senders must be knowledgeable about the data to be transmitted. An inadequate understanding of the basic data and/or a poor methodology for transmitting the information will almost certainly add to the likely inability of receivers to accurately respond to the message. Not surprisingly, the characteristics of the message and the nature of the senders interact to determine the quality of the final message received by the buyer. A complex message or a biased or selective transmission will have similar results.

Finally, there are some constraints which arise simply because of the context within which the message transmission occurs. A restricted housing market (few vacancies or few low rent units) will have major impacts on housing search and the receiver's willingness to filter and use new negative information.

## THE DISCLOSURE LEGISLATION

The original legislation was passed in March 1972 following the destructive San Fernando earthquake of February 1971. It required the state geologist to delineate by the end of 1973 "appropriate wide special studies zones to encompass all potentially and recently active traces of the San Andreas, Calaveras, Hayward, and San Jacinto Faults," as well as other faults which were a "potential hazard to structures from surface faulting or fault creep." These zones were to be one-quarter mile in width or less. The original legislation required that within these zones, city or county approval would be required for all new real estate development or structure for human occupancy, and it was specified that "cities and counties shall not approve the location of such a development or structure within a delineated special studies zone if an undue hazard would be created" (Section 2623, California Public Resources Code). The purpose of the act was thus to prevent new large-scale development or the siting of such facilities as hospitals and schools in areas particularly susceptible to fault rupture.

In 1975, a major series of amendments to the act were passed, including one mandating disclosure of the location of the special studies zones to persons considering the purchase of property within the zone. The disclosure amendment stated that "a person who is acting as an agent for a seller of real property which is located within a delineated special studies zone, or the seller if he is acting without an agent, shall disclose to any prospective purchaser the fact that the property is located within a delineated special studies zone" (Section 2621.9, California Public Resources Code). Given the strength of the real estate lobby in California, it might have been expected that the 1975 amendment would have generated controversy in the state legislature; instead, the act passed virtually unopposed after a few amendments were modified in the assembly.

Part of the reason for the acquiescence on the part of the California Association of Realtors was the package of amendments of which the disclosure provision was a part. It should be noted that along with disclosure, several changes favorable to real estate developers and agents were added, including a change of the name of the zones from "geologic hazard zones" to "special studies zones," the exemption of new single family frame dwellings which were not part of large developments from geologic reports, the exemption of mobile homes and condominium conversions from reports, and the exemption of alterations or additions to structures when such alterations do not exceed 50 percent of the value of the structure. Although the California Association of Realtors would have preferred that if disclosure was to be written into the law at all, it would be made the responsibility of the seller (rather than the real estate agent), they acquiesced to the language, given the rest of the package (Gillies, 1980). Proponents of the disclosure provision also viewed the final package of amendments as a compromise, in which they had traded the exemption of single family dwellings for the disclosure provision (Hurst, 1980).

After some initial confusion over the issue of how the real estate agent was to determine if a particular parcel was within or outside a special studies zone, and how, precisely, disclosure was to take place, a fairly standard procedure for disclosure was established. The standardization of disclosure was assisted by the 1977 publication of a well-written manual on special studies zone disclosure (California Association of Realtors, 1977), and the development of a contract addendum to the deposit receipt which was made available to California Realtors. Several Boards of Realtors took it upon themselves to produce colored maps outlining the location of the special studies zones (and sometimes other hazards areas such as flood-plains or landslide prone areas) which they either used in their offices, or gave to clients. The commission charged with regulating real estate practice, the California Department of Real Estate, seemed satisfied that disclosure was taking place -- from 1975 to 1978 there were only 13 disclosure complaints in all of northern

California, of which only two resulted in desist and refrain orders (Liberator, 1979).

## PRESENCE OF CONDITIONS FOR BUYER RESPONSE

Two hypotheses were generated from the list of conditions under which buyer response might be expected (Table 8.1). The first hypothesis was concerned with buyers and posed the question of the degree of importance of earthquake hazards. The second hypothesis focused on the agents and the accuracy and timing of the information presented to the buyers. Do real estate agents know that any given house is in a special studies zone, and can they define the meaning of this location in terms of probabilities of damage?

### Study Areas

Special study zones cover only a small part of the residential property in the State of California, although they are present in virtually every large metropolitan area (Figure 8.1). The zones include a wide range of property and social areas, from high-priced housing in the Berkeley hills and Portola Valley to lower cost housing in San Fernando and Antioch; they include areas which are inhabited by whites, Hispanics, and blacks; they include scenic areas with good views and high air quality, as well as areas of dense urban development and poor air quality (Grow and Palm, 1981).

The two study areas were chosen to minimize social and economic contrasts, but yet to vary in physical characteristics of site and Realtor organization. The study areas were Berkeley and central Contra Costa County. The neighborhoods within the special studies zones in both cases are white, and upper-middle class in socio-economic status. Both areas are suburban to San Francisco and are predominantly made up of single-family detached dwellings. The study areas differ in that they are located on different fault traces; Berkeley is on the Hayward fault, Contra Costa County is on the Calaveras fault. They have different amounts of visible damage from fault creep. The evidence of damage to houses, retaining walls, and curbs is far more apparent in Berkeley. In addition, the housing in Berkeley is older, individually designed, and has more brick and stucco. Finally, the two areas are served by different Boards of Realtors, with essentially independent submarket characteristics as defined by price-attribute structures and the probability of property exchange between board districts (Palm, 1978).

### Survey Design

Three populations were sampled in the survey research. The first was the set of homebuyers within the special studies zones in Berkeley and Contra Costa County which had purchased homes within the six months prior to the interviews. All recorded property

Table 8.1: Some of the Necessary Conditions for a Measurable Response to Negative Environmental Information

| | Characteristics of Buyer (Receiver) | Characteristics of Information (Message) | Characteristics of Real Estate Agents (Senders) | Setting |
|---|---|---|---|---|
| 1. | Buyer must consider environmental information salient to decision. | Information must be accurate and specific. | Agents must be knowledgeable about information they are transmitting. | Housing market must not be so restricted that there are actually few or no vacancies in other areas. |
| 2. | Buyer must feel that he has the latitude to act on information received<br>a. sufficient income to have alternative choices;<br>b. sufficient time to seek alternative choices;<br>c. knowledge of alternative destinations or knowledge of mitigation measures possible at destination. | Information must be clear and not contradicted with other stronger messages. | If agents have personal doubts about the significance of the message, they should not let these interfere with informing the buyers. | Adoption of avoidance or mitigation strategy by individual buyers must be economically rational in both the short- and the long-term view. |
| 3. | Variability in willingness to accept risk. | | | Reality of the message must be reinforced by observations by buyers -- visible damage or known future economic damage to investment. |
| 4. | Variability in fatalism. | | | |

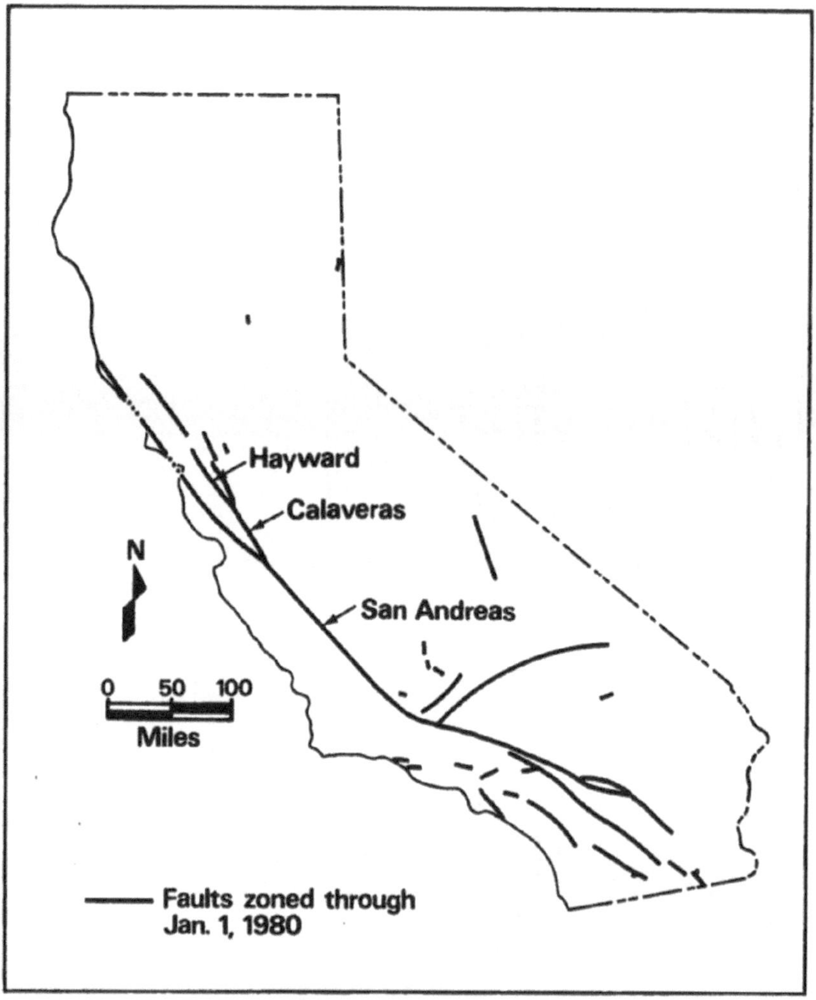


Figure 8.1. California special studies zones.

transfers within Alameda (Berkeley) and Contra Costa Counties as listed by the Registrar of Deeds were searched to compile a complete list of those who had purchased property within the special studies zones. Every household which had purchased a single family dwelling within the special studies zones over the period of August 1978 to January 1979 was contacted for an interview. The response rate was approximately 85 percent.

The second population was a set of homebuyers which had purchased houses near, but not actually within the special studies zones. This set of homebuyers was surveyed to determine if there is a difference in attitudes between those who move into special studies zones despite disclosure, and those who move elsewhere. It was expected that this population might include those who had been discouraged from moving within the special studies zones by the disclosure, and had chosen to live outside the zone. Using the same records, a set of 50 homebuyers from each study area who had purchased houses within five miles of the special studies zones and within neighborhoods of similar physical and social characteristics were selected, beginning with those who had purchased houses in August 1978 until the quota was met.

The third population was the set of real estate agents who were identified by asking the buyers to name the real estate agent who had "helped them with their home purchase." It was felt that any other sampling might include persons merely holding real estate licenses but who did not actually sell real estate, or agents not familiar with the disclosure because no special studies zone was within their sales territory. The response rate from this population was about 90 percent.

## RESULTS I -- THE IMPORTANCE OF EARTHQUAKE HAZARDS

In general, the homebuyers surveyed did not attach much importance to factors within the physical environment in their decision to buy a house. Each homebuyer interviewed was asked to indicate the importance of a set of fifteen factors with respect to the purchase decision. In both study areas, distance from an active earthquake fault, and location outside a floodplain were ranked as the least important in the purchase decision. Most important in both areas were investment potential, price, beauty of the area, and size of the dwelling unit. In Berkeley, 79 percent and in Contra Costa County, 85 percent said that distance from an active earthquake fault was not important or was not even considered in the purchase decision (Table 8.2). This lack of concern with distance from an active earthquake fault was expressed both by buyers within the special studies zones and those who had located in nearby areas outside the special studies zones (Table 8.3).

Homebuyers were asked whether the location of the special studies zones made any difference in their decision to buy this particular house. Since not all homebuyers within the special studies zones were aware of their location, this question was posed only to those who could remember the disclosure and who could properly identify that the disclosure was related to proximity to an active earthquake fault. There was no significant difference in the response to this question by homebuyers either within or outside the special studies zones: the location of the house within or outside the zone generally did not make any difference in the purchase decision (Table 8.4).

Table 8.2: Housing Factor Ratings with Respect to the Decision to Buy the Present Home

| Factor | | Very Important | Somewhat Important | Not Important | Did Not Consider |
|---|---|---|---|---|---|
| Investment potential or resale | B | 25 | 9 | 4 | 1 |
| | C | 131 | 22 | 3 | 1 |
| Price | B | 23 | 15 | 1 | 0 |
| | C | 110 | 50 | 6 | 0 |
| Beauty of the area | B | 24 | 13 | 1 | 1 |
| | C | 82 | 57 | 6 | 1 |
| Number of bedrooms | B | 18 | 13 | 6 | 2 |
| | C | 71 | 70 | 25 | 0 |
| Views | B | 15 | 21 | 2 | 1 |
| | C | 53 | 83 | 26 | 1 |
| Distance to work | B | 13 | 17 | 7 | 2 |
| | C | 56 | 62 | 39 | 9 |
| Social composition of the neighborhood | B | 11 | 20 | 5 | 2 |
| | C | 49 | 74 | 31 | 12 |
| Reputation of crime rate in the n'hood | B | 11 | 15 | 5 | 8 |
| | C | 51 | 71 | 30 | 13 |
| Air quality | B | 5 | 13 | 9 | 12 |
| | C | 57 | 53 | 29 | 27 |
| Quality of local public schools | B | 5 | 10 | 14 | 10 |
| | C | 59 | 37 | 38 | 32 |
| Closeness to school | B | 7 | 8 | 17 | 7 |
| | C | 49 | 40 | 46 | 31 |
| Closeness to friends or relatives | B | 11 | 12 | 10 | 6 |
| | C | 16 | 45 | 64 | 21 |
| Accessibility to public transportation | B | 2 | 10 | 13 | 14 |
| | C | 13 | 26 | 81 | 46 |
| Distance from active earthquake fault | B | 2 | 6 | 20 | 11 |
| | C | 14 | 23 | 63 | 66 |
| Location out of a floodplain | B | 0 | 4 | 8 | 27 |
| | C | 21 | 35 | 36 | 73 |

*Note:* B = Berkeley, n = 39 and C = Contra Costa, n = 166.

Table 8.3: Comparison of Buyers Within Special Studies Zones and Adjacent to Special Studies Zones (Evaluation of "Distance from an Active Earthquake Fault" in Importance in Decision to Buy Your Home).

| | Adjacent[1] Berkeley | Inzone[2] Berkeley | Adjacent Contra Costa | Inzone Contra Costa |
|---|---|---|---|---|
| Very important | 3 | 2 | 2 | 14 |
| Somewhat important | 13 | 6 | 5 | 23 |
| Not important | 10 | 20 | 10 | 63 |
| Did not consider | 9 | 11 | 24 | 66 |
| | (<.90/no significant difference) | | (<.01/no significant difference) | |
| Have you ever heard the term "special studies zone or Alquist-Piolo zone?" | | | | |
| Yes | 28 | 34 | 13 | 45 |
| No | 11 | 7 | 25 | 120 |
| | (.75/no significant difference) | | (.57/no significant difference) | |

[1]"Adjacent" was defined as within five miles of a special studies zone in a comparable neighborhood.

[2]"Inzone" was a resident living within the special studies zone.

Most residents (85 percent) of the special studies zones felt this location would not affect the price of the house or the ability to sell it when they decided to move. However, there was a difference between the opinion of the special studies zones residents and those who lived nearby as to the likelihood of losses to those who lived in the zones. Residents of the zones generally felt there was no difference in potential losses, with only 29 percent indicating that they were more susceptible to losses from earthquakes. Residents of nearby areas were more likely to feel that those living in the zones are more susceptible to losses from earthquakes (61 percent).

In sum, homebuyers generally attach little importance to earthquake hazards or proximity to an active earthquake fault in the home purchase decision. Furthermore, they are generally not convinced that special zones actually outline areas particularly

Table 8.4: Responses to the Question -- Did the Location of Earthquake Hazard Zones Make Any Difference in Your Decision to Buy this Particular House? (for those who knew they were in a special studies zone)

| | Adjacent[1] Total | Inzone[2] Total |
|---|---|---|
| Yes | 15 | 18 |
| No | 57 | 78 |
| | (.11/no significant difference) | |

| | Adjacent Berkeley | Inzone Berkeley | Adjacent Contra Costa | Inzone Contra Costa |
|---|---|---|---|---|
| Yes | 12 | 9 | 3 | 9 |
| No | 23 | 28 | 34 | 50 |
| | (.48/no significant difference) | | (.23/no significant difference) | |

[1]"Adjacent" was defined as within five miles of a special studies zone in a comparable neighborhood.

[2]"Inzone" was a resident living within the special studies zone.

susceptible to damage. This factor in and of itself should reduce the impact of disclosure legislation on buyer response.

## RESULTS II -- THE QUALITY AND TIMING OF INFORMATION

It has been argued that even if buyers were concerned with earthquake hazards information, it would be necessary that such information be of high quality and presented at an appropriate time in the purchase decision for it to be translated into a measurable response. In the case of the disclosure of special studies zones locations, it was found that in many cases this condition was lacking.

First, although most of the real estate agents interviewed could associate the term "special studies zones" with earthquake hazards, 14 percent of the agents misidentified the term. It is important to recall that these are persons mandated by the law to disclose that property is within the special studies zones, and that the real estate agents interviewed were those who had recently sold houses within these zones. If the real estate agents were not fully aware of the meaning of the special studies zones, it is not surprising that buyers did not understand what the zones meant. Indeed, only 95 of the total 207 of the interviewed homebuyers living within special studies zones could either identify the meaning of the term "special studies zone" or recall any kind of disclosure related to the location of the house.

Second, real estate agents are not convinced that the special studies zones are particularly important or meaningful. Only about a third believe that people living within the special studies zones are more likely to suffer financial losses or physical injuries in the event of an earthquake. Given these beliefs, the real estate agent can comply with the disclosure law, but minimize the psychological impact of the disclosure by downplaying its importance to his clients. Several statements from the real estate agents, including: "This is just another government regulation" and "We don't get earthquakes in Contra Costa County" are examples of mixed messages from the information sources -- the agents. In sum, because many real estate agents are sincerely not convinced that special studies zones demarcate areas of particular earthquake hazard, they downplay the role of hazard information. (Of course, it should be noted that the business interests of the real estate agents would be hampered if buyers were to take disclosure too seriously.)

Finally, standard disclosure procedures minimize the impact of disclosure through both timing and method. Timing of disclosure is particularly important because it can have a strong effect on the purchase decision. Previous research has shown that the timing of the presentation of information on vacancies can affect both the length of search and the purchase decision (Clark and Smith, 1979; Smith and Mertz, 1980). In the case where negative information must be presented, it is an accepted sales procedure to introduce such information after commitment to the purchase has already been shown by the prospective buyer and the household has decided how it will finance the purchase. This moment is at the time the purchase contract is being signed. On the other hand, the most sensitive time for information presentation is when the house is actually being shown, and the prospective buyers are carefully noting the positive and negative aspects of the house. Negative information presented at this time can have a magnified influence on the purchase decision. If real estate agents were to manipulate the timing of disclosure to their best advantage, they would disclose at the time of the signing of the purchase contract, and would avoid disclosure when the house was actually being shown to the client. This is precisely the current practice. When real estate agents were asked when they routinely make disclosure, 90 percent indicate that they disclosed at the contract signing time, and only 9 percent at the time the house was being shown. Thirty-two percent also indicated they introduced the buyer to special studies zones before any houses were shown, during the office interview. (Percentages do not add to 100 because some real estate agents answered that they disclose special studies zone location at more than one time during the sales process.)

Method of disclosure also serves to minimize its impact on the buyer. The three standard formats used for disclosure are the information in the Multiple Listing Service book (used by 30 percent), a map of the area (used by 70 percent), and the contract addendum (used by 90 percent of the respondents). The Multiple Listing Service book form presents little information to the homebuyer. In Berkeley, the disclosure on this form is merely a typed line stating "in Alquist-Priolo zone" or even "in Alquist-Priolo district." Surely, to the uninitiated buyer, such a statement might mean anything, and would most probably indicate the names of the state legislators for the district rather than anything to do with earthquake hazards. In Contra Costa County, the form includes a line stating "special studies zone" and a box which can be marked "yes" or "no". Again, such a disclosure would tell the buyer very little.

The map, used particularly in Contra Costa County, is a detailed street map of the region with the one-hundred year floodplain in blue, the special studies zone in yellow, and the combination of areas in green. Again, the terms are not defined on the map. Finally, the most frequently used method of disclosure is a contract addendum which until recently has stated that "the property is or may be situated in a Special Study Zone." No further definition of the special studies zone is presented, but the form notes that although construction for human occupancy may be subject to the findings of a geologic report, single family woodframe dwellings are exempted, as are structures in existence prior to May 4, 1975. The words "seismic," "hazard," "earthquake," or "fault" are not included in the contract addendum.

The analysis of the accuracy and timing of disclosure indicates a further barrier in the way of potential information impact. When inaccurate information is combined with biased timing, it is unlikely that there will be little influence on buyer behavior.

## THE IMPACTS OF DISCLOSURE ON HOMEBUYERS

Three indices are used to measure potential impact of disclosure on home buyers: measures of the frequency of buyers refusing to purchase a property based on disclosure of special studies zones location, the effect of special studies zones locations on house prices, and the mitigation measures taken by special studies zones residents.

Most real estate agents reported that they had never had a client decide to forego purchase of a home after being informed that the property was in a special studies zone. Only 12 of 74 agents had ever experienced a refusal, and of these only two had more than three prospective clients decline to purchase a house in a special studies zone. Responses to the question, "Have you ever had a client decide not to buy a home after being informed that the property was in an Alquist-Priolo Special Studies Zone," were cross-tabulated with the timing and methods used in disclosure. The only statistically significant effect was the use of the Board of Realtors map, and this effect was counter-intuitive: agents who used the map had fewer refusals than those who did not. One can only conclude that the presentation of the map convinced buyers that because many other properties were in the same situation, the special studies zones were not cause for particular concern. Other variations in timing or materials used did not make a difference in the likelihood of clients refusing to purchase the house.

House prices were not affected by special studies zones locations. To test for the effects of special studies zones on house prices, hedonic price indices were calculated on house price levels in 1972, before the disclosure legislation was in effect, and in 1977, after the disclosure was in place. Data on properties sold were obtained from the appraisal reports filed with the Society of Real Estate Appraisers, and included such descriptions of the property as square footage of dwelling space, age of house, quality of house, condition of house, size of lot, and presence of such amenities as view lot, swimming pool, and fireplaces. In addition, data from the 1970 Census of Population on economic status of the area (percentage professional-managerial in the tract), and housing composition (percentage of single family dwellings in the tract) were added. Location in the special studies zone was added at the second step in the ordinary least squares equation as a dummy variable: the property was considered to be within the special studies zone, close to (within five miles of) the special studies zone, or outside (beyond five miles). The research hypotheses tested were that: in 1972, location in the special studies zone should be unrelated to house price, and the regression coefficient for the dummy variable should be close to zero; in 1977, location in a special studies zone should have a statistically significant negative regression coefficient, location near the special studies zone should have a statistically significant positive regression coefficient (because of a build-up of demand in proximate areas outside the zone itself), and location outside the zones should have no effect on house prices.

Results of the ordinary least squares analysis show that in Berkeley, location in the special studies zone had a positive effect on house prices in both years, location near the zones had a positive effect on house prices, and locations outside the zones had a negative effect on prices in both years (Table 8.5). These findings are almost completely opposite the ones predicted for the research hypotheses. Although one would not argue that special studies zones are actually viewed as an advantage by Berkeley homebuyers, it is safe to say that the lower hills and Claremont, in which the special studies zones lie, add value to the house which is not sufficiently accounted for by variables such as "view" or overall social status of the neighborhood. At the very least, one must conclude that special studies zonation had no depressing effect on house prices within Berkeley.

Table 8.5: Hedonic Price Index for Effects of Special Studies Zone Location on House Prices in 1972 and 1977

| | B | Statistical Significance | Multiple $R^2$ |
|---|---|---|---|
| Berkeley | | | |
| Within special studies zone | | | |
| 1972 | $ 1,607 | .97 | .87 |
| 1977 | 7,318 | .99 | .65 |
| Up to 5 miles from special studies zone | | | |
| 1972 | 494 | .57 | .86 |
| 1977 | 8,801 | .99 | .64 |
| More than 5 miles from special studies zone | | | |
| 1972 | -2,001 | .99 | .87 |
| 1977 | -11,548 | .99 | .64 |
| Contra Costa County | | | |
| Within special studies zone | | | |
| 1972 | -853 | .93 | .83 |
| 1977 | -4,867 | .99 | .77 |
| Up to 5 miles from special studies zone | | | |
| 1972 | -528 | .71 | .83 |
| 1977 | -1,450 | .97 | .77 |
| More than 5 miles from special studies zone | | | |
| 1972 | 840 | .97 | .83 |
| 1977 | 2,234 | .94 | .78 |

Variables Used:

square footage of dwelling unit
lot size
condition of dwelling unit
quality of dwelling unit
age of house
"view"
presence of fireplace
presence of swimming pool
type of mortgage (conventional or insured)
percentage professional and managerial in census tract
percentage single family dwellings in census tract

In Contra Costa County quite different results were obtained. Here location in the special studies zones had a negative effect on house prices in both years, locations near the zones took on a statistically significant positive effect, and locations distant from the zones maintained a positive effect on house prices. Because the direction of the relationship between location and house prices did not change for the special studies zones, one must also conclude that the zonation itself did not dampen house prices here, but that rather, the negative effect of the special studies zone location was due to other factors of longer standing -- perhaps proximity to the freeway which has been placed almost astride the Calaveras fault.

Finally, few residents of the special studies zones reported that they had taken measures to mitigate against damage from a possible future earthquake. To obtain information on mitigation measures adopted, a mail resurvey was conducted of only those residents of the special studies zones who were aware of their location with respect to the fault trace. Fifty-eight of 97 such households responded, or a response rate of about 60 percent. The responses of these households were compared with a survey of 1,450 Los Angeles County residents (Turner, Nigg, Paz and Young, 1979). Recent buyers were asked to indicate for a series of mitigation measures, whether they planned to take a mitigation strategy or had no such plans and whether they had taken the strategy because of the threat of earthquake damage or for other reasons. Mitigation measures ranged from such common items as the possession of a working flashlight and battery radio, to more complex procedures for family or neighborhood emergency plans. It should be noted that the Los Angeles sample was a random sample of the entire population, and was not limited to those living within special studies zones or who had received any formal disclosure of proximity to an active fault trace.

A comparison of the responses of the two populations (Tables 8.6 and 8.7) shows two facts: first, only a small percentage of either population has taken any kind of mitigation measure because of the threat of earthquakes, and second, in most cases the southern California sample has taken more mitigation measures than the sample within the special studies zones in the northern California study areas. A far larger percentage of Los Angeles residents had stored food and water, had instructed children what to do in an earthquake and made plans for a family reunion, and had set up neighborhood responsibility plans. The Bay Area residents exceeded the Los Angeles residents only in replacement of cupboard latches, structural reinforcements of dwelling units, and the investigation or purchase of earthquake insurance. Even these mitigation measures were taken by only a small minority of those who lived in and were aware of the significance of special studies zones.

To summarize, disclosure had little measurable effect on buyer behavior. Few real estate agents reported that buyers balked at a home purchase simply as a result of the disclosure of special studies zone location. The special studies zones had no measurable impact on house prices, indicating that there was no lessening of overall

Table 8.6: Percent of Respondents Who Have Taken Specific Preparation Measures for Earthquakes in Los Angeles

| Mitigation Measure | For Earthquakes | For Other Reasons | Plan to Do | Don't Plan |
|---|---|---|---|---|
| Working flashlight | 10.8 | 60.7 | 16.6 | 11.9 |
| Working battery radio | 11.1 | 43.5 | 17.5 | 27.9 |
| First aid kit | 8.0 | 46.1 | 18.6 | 27.0 |
| Store food | 8.0 | 18.8 | 17.4 | 55.8 |
| Store water | 8.0 | 9.1 | 15.2 | 67.7 |
| Rearrange cupboard contents | 9.7 | 6.6 | 8.1 | 76.2 |
| Replace cupboard latches | 4.5 | 5.7 | 6.5 | 83.3 |
| Contact neighbors for information | 9.8 | 9.7 | --- | 80.5 |
| Set up neighborhood plans | 4.0 | 8.2 | --- | 87.8 |
| Consider earthquake insurance | 23.1 | 6.7 | --- | 70.2 |
| Bought earthquake insurance | 12.8 | 5.2 | --- | 82.0 |
| Structurally reinforced home | 4.7 | 6.4 | 4.8 | 84.1 |
| Instruct children what to do in earthquake | 47.6 | 2.8 | 20.8 | 28.8 |
| Family plans for reunion after earthquake | 19.9 | 2.2 | 15.7 | 62.4 |
| Emergency procedures at residence | 26.1 | 8.0 | 21.7 | 44.2 |

*Source:* Los Angeles data summarized from Ralph H. Turner et al., Earthquake Threat: The Human Response in Southern California (Los Angeles: Institute for Social Science Research, University of California, Los Angeles, 1979).

Table 8.7: Percent of Respondents Who Have Taken Specific Preparation Measures for Earthquakes in San Francisco

| Mitigation Measure | For Earthquakes | For Other Reasons | Plan to Do | Don't Plan |
|---|---|---|---|---|
| Working flashlight | 6.9 | 79.3 | 12.1 | 2.7 |
| Working battery radio | 8.6 | 44.8 | 25.9 | 20.7 |
| First aid kit | 3.4 | 65.5 | 22.6 | 8.6 |
| Store food | 1.7 | 19.0 | 20.7 | 58.6 |
| Store water | 1.7 | 3.4 | 25.9 | 68.9 |
| Rearrange cupboard contents | 5.2 | 6.9 | 8.6 | 79.3 |
| Replace cupboard latches | 3.8 | 8.6 | --- | 77.6 |
| Contact neighbors for information | 3.4 | 12.1 | --- | 84.5 |
| Set up neighborhood plans | 1.7 | 10.3 | --- | 88.0 |
| Consider earthquake insurance | 41.4 | 17.2 | --- | 41.4 |
| Bought earthquake insurance | 24.1 | 10.3 | --- | 65.5 |
| Structurally reinforced home | 8.6 | 5.2 | --- | 86.2 |
| Instruct children what to do in earthquake | 20.0 | 2.2 | 46.6 | 31.2 |
| Family plans for reunion after earthquake | 14.0 | 2.0 | 44.0 | 40.0 |
| Emergency procedures at residence | 15.6 | 9.8 | 60.8 | 13.7 |

*Source:* San Francisco data based on mail survey by author to respondents in Berkeley and Contra Costa County who had indicated they were aware they lived in a Special Studies Zone.

demand within the zones as compared to other areas. Finally, few residents of the special studies zones have bothered to take mitigation measures, and most have no plans to do so in the foreseeable future. Disclosure of a message with low salience, and with little attempt to persuade the recipients of its importance has resulted in little or no measurable impact on the intended receivers.

## DISCUSSION

The overall negative results of this study of information and information provision raise serious questions about our efforts to influence and modify search and relocation behavior. And, given the effort which federal and state governmental bodies have taken to provide information to consumers about environmental hazards associated with site, the costs of drafting and updating maps, and the costs to both the government and the real estate industry to administer and enforce the existing program, one must feel disconcerted that all of this effort seems to result in no measurable change in behavior on the part of the homebuyers. Aside from the obvious problems of the need for greater accuracy in message transmission on the part of the real estate agents, there seem to be three factors which are presently interfering with the translation of special studies zone information into change in buyer behavior: the primary motivation of buyers, the definition of the source information (the zones), and the communication process.

As the buyers themselves have indicated, the primary motivation which homebuyers have is to minimize the price they pay for a dwelling unit of given characteristics, and to maximize the potential resale value. In short, economic motivations are primary in the purchase decision. Since the house is viewed as an economic investment, consistent with the notions put forward by Perrin (1977) about the social values of housing and neighborhood in the United States, buyers do not see themselves as making a long-term commitment either to the house or to the neighborhood. Buyers expect to stay in the house for a relatively short time, and many studies have shown that they will move within three to five years, depending on the metropolitan setting, as well as the stage in the life cycle of the buyers.

Since buyers believe that a major damaging earthquake will not occur in the three to five years in which they are living in the house, a belief which is not unreasonable, they do not worry about buying a home as long as it has a good potential resale value. They will not see it as economically rational to take costly mitigation measures such as structural reinforcements or even the purchase of earthquake insurance with a high deductible, since these expenses cannot be recouped in a subsequent house sale in the same way as investment in an additional bathroom or patio can. The short-term decision to forego mitigation measures, and proceed with the purchase of a house in a hazardous area must therefore be understood as economically rational from the point of view of the individual homeowner, as unpleasant as this realization may be to policy-makers or those viewing potentials for disaster at a community-wide level. Unless hazards and the information about hazards have economic impacts, then it is unlikely the information will have any measurable impact.

The special studies zones themselves may also lead the unusually well-informed homebuyer to ignore their consequences. As was reviewed earlier, disclosure of the zones was passed as a compromise measure in exchange for an agreement to forego the construction of large-scale residential and other projects astride active fault traces. However, it must be noted that the zones do not outline all of the areas susceptible to earthquake damage, and indeed may include areas safer than those lying outside the zones. (It is of interest that no zone runs through the city of San Francisco.) The special studies zones do not include areas susceptible to damage from liquification, shaking, or ground failure, but rather, only those susceptible to damage from rupture and fault creep. This is because the zones are not based on bedrock conditions, but merely on simple distance from surface fault traces. The sophisticated buyer, who may know that although his house is near a fault but not directly on one, is actually correct in his assessment that purchase of a house in a special studies zone does not mean that he is necessarily more liable to injury or property damage in the event of a major earthquake. Although few buyers actually contract for a study of the position of their individual property with respect to the fault, it is possible that some buyers may be quite right in their assessment of the fact that their house may be less susceptible to damage than those built on unstable slopes or landfill. This raises the critical issue of the accuracy of the source information as well as the problem of the accuracy of presentation.

Finally, if county and state governments did provide maps of hazardous areas, and if it were the goal of legislation to effectively transmit this information to prospective homebuyers, the communication process itself should be reconsidered. Communication theorists emphasize the need for multiple and reinforcing messages if persuasion is to be effective (Rogers and Ararwala-Rogers, 1976). It is not enough for information to be passed to homebuyers in a single contract addendum by an information agent unsympathetic to persuading the buyer to give heed to the message; rather, the message needs to be reinforced. Some possible methods of reinforcement would include broader public-awareness programs, as well as specific information systems aimed at homebuyers through the appraisal and title search process. The provision of information alone is insufficient to evoke a change in behavior and impact household search and relocation behavior.

## REFERENCES


California Association of Realtors (1977) Disclosure of Geologic Hazards. Los Angeles: California Association of Realtors.

Clark, W. A. V. and T. R. Smith (1979) "Modelling information use in a spatial context", Annals of the Association of American Geographers 69, 575-588.

Gillies, D. (1980) Vice President, Governmental Relations, California Association of Realtors. Personal communication.

Grow, C. and R. Palm (1981) Population and Housing in the Special Studies Zones. Boulder, Colorado: University of Colorado, Institute of Behavioral Science, Occasional Paper No. 1.

Hurst, B. (1980) Former Legislative Assistant to State Senator Alfred Alquist. Personal communication.

Kockelman, W. J. (1980) Examples of the Use of Earth-science Information by Decisionmakers in the San Francisco Bay Region, California. U.S. Geological Survey Open-File Report No. 80-124.

Liberator, J. (1979) California Department of Real Estate. Personal communication.

McCarthy, K. (1979) "Housing search and residential mobility", Washington, D.C.: Paper presented at the Conference on the Housing Choice of Low Income Families.

Palm, R. (1976) Urban Social Geography from the Perspective of the Real Estate Salesman. Berkeley: University of California, Center for Real Estate and Urban Economics.

----- (1978) "Spatial segmentation of the urban housing market, Economic Geography, 54, 210-221.

Perrin, C. (1977) Everything in Its Place: Social Order and Land Use in America. Princeton, New Jersey: Princeton University Press.

Rogers, E. M. and R. Ararwala-Rogers (1976) Communication in Organizations. New York: The Free Press.

Smith, T. R. and W. A. V. Clark (1980) "Housing market search: information constraints and efficiency", in W. A. V. Clark and E. G. Moore Residential Mobility and Public Policy. Beverly Hills, California: Sage Publications.

----- and F. Mertz (1980) "An Analysis of the Effects of Information Revision on the Outcome of Housing-market Search, with special reference to the influence of realty-agents", Environment and Planning A, 12, 155-74.

Turner, R., J. Nigg, D. Pas, and B. Young (1979) Earthquake Threat: The Human Response in Southern California. Los Angeles: University of California, Institute for Social Science Research.

# 9

# LINKING LOCAL MOBILITY RATES TO MIGRATION RATES: REPEAT MOVERS AND PLACE EFFECTS

**John L. Goodman, Jr.**

It is now an established practice in studying local residential mobility to treat residential change as a two stage process. First the individual or household decides to move. Then, given that they are moving, they choose the destination. Substantial evidence indicates that the determinants of moving differ from the determinants of the destination choice (Speare, Goldstein, and Frey, 1975, and references therein).

As applied to analysis of out-movement from cities to their suburban ring, this literature argues that the mobility propensity of city populations should be analyzed separately from the city/ring destination choice of city movers. Frey (1979) has shown that both mobility propensity differentials and destination choice differentials across metropolitan areas contribute to the observed variation across metropolitan areas in rates of white city-to-suburb mobility.

With regard to the first decision, the decision to move, previous research indicates that household life cycle changes and the accompanying need for housing adjustments are foremost among the causes (Rossi, 1955; Goodman, 1979). Neighborhood attributes are not important precipitators of moves.

Still, when local mobility rates are computed for individual SMSAs, substantial variation is observed (Figure 9.1). The determinants of this variation in local mobility rates across SMSAs have not previously been rigorously examined. The tone of the literature is, however, clearly that high local mobility rates are attributable to characteristics of the area's residents. San Jose is implied to have a higher local mobility rate than Cleveland because the former has a higher proportion of "mobility prone" individuals, that is, individuals whose mobility history, life cycle stage, and other demographic characteristics are associated with high mobility propensity.

*AUTHOR'S NOTE: This research was supported in part by the Center for Population Research, National Institute of Child Health and Human Development, through grant HD 13442-01. Opinions expressed here are those of the author and do not necessarily represent the views of The Urban Institute or its sponsors.*

DOI: 10.4324/9781003182085-11 

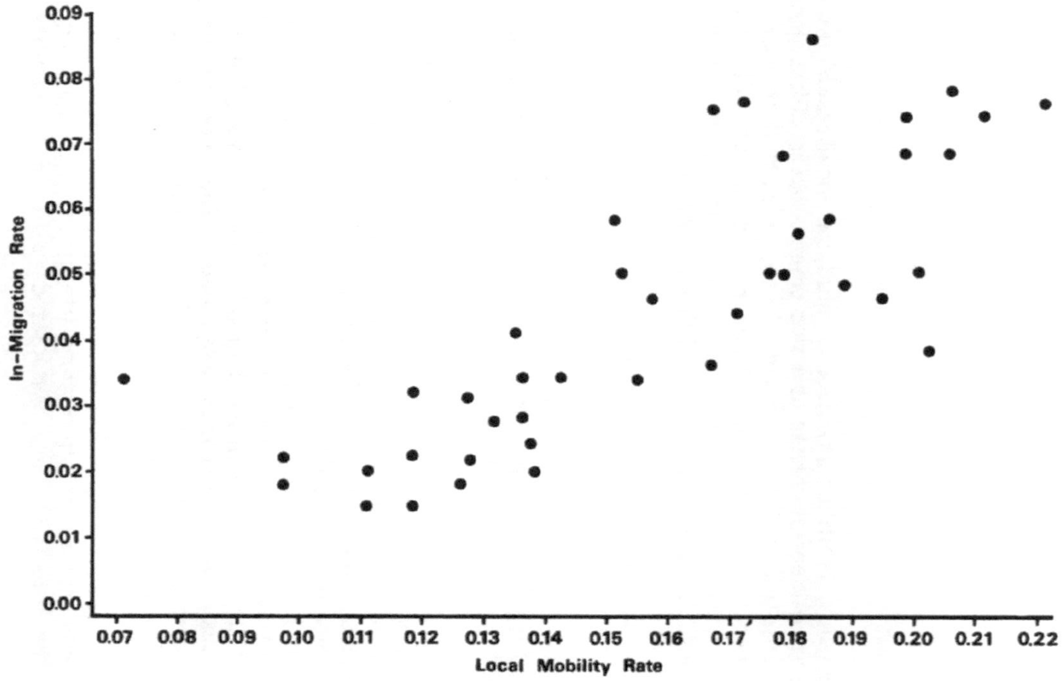


Figure 9.1. In-migration rates and local mobility rates for 42 SMSAs. *Source:* Annual Housing Surveys (18 SMSAs excluded due to suppressed location codes). *Definitions:* In-migration rate: Annual number of household heads moving into the SMSA as a proportion of all households in SMSA. Local mobility rate: Annual number of household heads moving within SMSA as a proportion of all households in SMSA.

Findings from the recently completed housing allowance experiments raise doubts about the adequacy of this household-based explanation for spatial variation in local mobility rates. In these experiments, many households had to move to higher quality housing in order to receive the housing subsidy. Households at the Phoenix site were more successful than households in Pittsburgh in becoming recipients, largely because of their higher mobility rates. Furthermore, the higher mobility rates in Phoenix could not be explained by site variability in households' socioeconomic characteristics and past mobility behavior. "There is something over and above these (household-based) factors that leads to a higher mobility rate in Phoenix" (MacMillan, 1980). This remained an unresolved question in the housing allowance analysis.

The purpose of this paper is to explain variation across metropolitan areas in local residential mobility rates, by analyzing the correlation between the local mobility rate in an area and the in-migration rate to that area. Specifically, I develop and estimate a model in which local mobility rates are causally dependent on in-migration rates to the area. Use of micro-level data allows me to decompose the correlation observed at the SMSA level into components representing distinct causal connections between the two rates.

Previous work has noted the correlation between SMSA in-migration and local mobility rates (Long and Boertlein, 1977). In Figure 9.1 for example, the simple correlation (r) between the two rates is .79. But to date there has been no empirical analysis of this linkage. Previous theoretical work can, however, generate testable propositions, and we turn now to those theories.

## THE REPEAT MOVER THEORY

The first explanation for the correlation between migration rates and local mobility rates is the "repeat mover" theory, which contends that multiple moves by the same individuals are responsible for the macro-level correlations. Originally formulated to explain the positive correlation between in- and out-migration rates, the repeat mover theory states that many of the out-migrants from an area during the current time period were in-migrants during the previous time period (Miller, 1967; Morrison and Relles, 1975). The argument is that the same personal characteristics that lead people to migrate once will lead them to migrate again. In addition, unfulfilled expectations after a first move induce some migrants to migrate again (DaVanzo, 1978).

As applied to the linkage between migration and local mobility, the repeat mover theory has been articulated best by Roseman (1971). Because less information about the destination area is available to long-distance migrants ("total displacement" movers in Roseman's framework) than to local ("partial displacement") movers, Roseman hypothesized that the former would be less likely to initially secure appropriate housing and therefore would be more likely to make a subsequent local move.

Migrants would be expected to have higher subsequent mobility rates than local movers for several reasons. Migrants make housing choices based on their exposure to only a limited number of specific vacancies as well as only partial knowledge of the community and neighborhood alternatives that are available. Because of the distance between origin and destination, housing search is especially costly for long-distance movers. Also, some long-distance moves--such as corporate transfers--are made on short notice, further limiting the amount of housing search that can take place prior to housing selection. Unless the migrants have previously lived in the area to which they are moving, they will have little knowledge of the communities and neighborhoods from which to choose.

Migrants often are uncertain as to the housing they want and can afford, and this uncertainty also affects the initial housing selection. Households who move to an area to look for a job have to select housing without knowing how much they will eventually be able to afford for housing or the location(s) of their employment. After employment is secured, the initial housing may no longer match the household's financial capabilities or may be inconvenient to the work site.

Compared to migrants, local movers have more knowledge about local housing market conditions, more flexibility in scheduling their move, and thus the ability to wait for their "dream house" to appear on the market. In addition, since local moves are not as highly correlated with employment changes as are long-distance moves, local movers probably have better knowledge of their future income and employment location.

For reasons then of both housing market information and uncertainty as to housing needs, migrants might be expected to initially select temporary housing upon moving into an area and therefore to have a high rate of subsequent local mobility as they move into more permanent quarters. Their subsequent mobility is due not only to their previous move, but also to the distance of that move.

## THE PLACE-EFFECT THEORY

A second micro explanation for the macro correlations between in- and out-migration rates and between long-distance and local rates is the "place-effect" theory. It argues that in-migration to an area changes that area in a way that induces non-immigrants to leave the area. As Stone explained in his analysis of in- and out-migration:

> The correlation between in-migration and out-migration may also be partly a result of areal variation in social and economic changes. Areas undergoing relatively rapid change may generate conditions that attract in-migrants and at the

> same time repel former residents. Of course, migration itself may be involved in social and economic changes, as where an area becomes subject to a significant influx of persons from a relatively strange culture. This initial influx can set off a whole train of socioeconomic changes that trigger further in-migration while at the same time precipitating out-migration of former residents. (Stone, 1971:700)

As applied in the connection between in-migration and local mobility, the place-effect theory holds that in-migration increases the local mobility of current residents by altering neighborhood composition and housing market conditions. "It seems reasonable to suppose that a high volume of in-migration to a city would alter the composition of many neighborhoods and in the process generate 'locational stress.' Hence, one would expect to find a positive correlation between in-migration rates and rates of intra-city residential mobility" (Long and Boertlein, 1977). Furthermore, to the extent that friendships with neighbors inhibit residential mobility, areas with high rates of in- and out-migration offer less opportunity for these mobility-inhibiting friendships to be established.

Additional place effects operate through the housing market. Housing market conditions are influenced by in-migration, and may in turn affect the local mobility behavior of current residents. Because of both new construction and high turnover of existing housing units, more housing units come on the market in growing areas than in otherwise comparable areas that are not growing. Therefore, residents of growing metropolitan areas are exposed to a larger number of alternative housing opportunities. Other things equal, the more housing alternatives one knows about, the greater the probability that one will be judged superior to the current residence.

The chronic mover theory and the place-effect theory do not contradict each other. Both may be true, and both may contribute to the observed correlation between a metropolitan area's in-migration rate and its local mobility rate.

## A MICRO-LEVEL MODEL OF TENURE CHOICE AND SUBSEQUENT LOCAL MOBILITY

We need a micro-level model to test the theories described above and thereby to decompose the macro correlations into specific causal links. Our model (Figure 9.2) has two equations. Tenure choice is the dependent variable in the first equation, and subsequent local mobility is the dependent variable in the second equation.

Looking first at the determinants of movers' tenure choice, life-cycle stage and income have been shown in previous studies to be theoretically plausible and empirically strong determinants. Pre-move housing tenure is expected to affect movers' tenure choice in two ways. First, the proceeds from the sale of one home can be

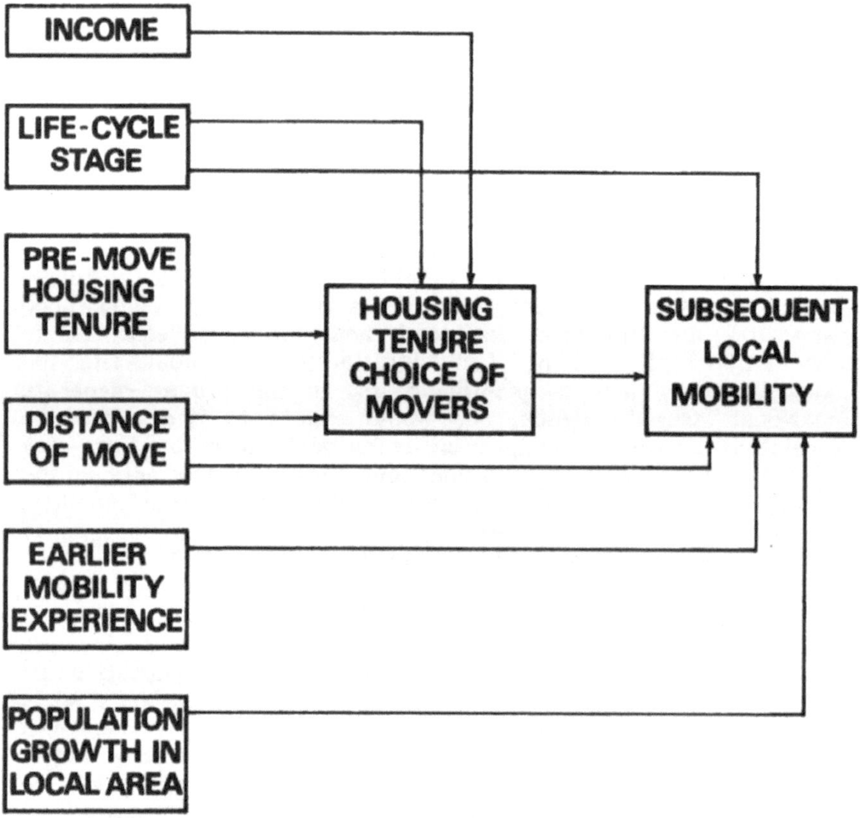


Figure 9.2
The model of tenure choice and subsequent local mobility.

used as a down payment on the purchase of the next. Second, pre-move housing tenure may indicate preferences for homeownership that are not captured by life-cycle stage. Previous research supports both suppositions (Weicher, 1977; McCarthy, 1976).

The fourth variable in the tenure choice equation, distance of move, is the determinant of special interest here. Our expectation is, other things equal, that:

Hypothesis 1. The longer the distance of move, the lower the probability of home purchase.

Households moving into an area generally have less time for search, less prior knowledge of the housing market, and less certainty that they will be staying in the area. All of these factors should lead the in-migrants to exercise caution and avoid risk in their initial selection of housing. Since renting involves much lower transaction costs than does purchase or sale of a house, in-migrants should be less likely to buy, and more likely to rent, than otherwise comparable local movers.

The second equation in the model specifies the probability of a subsequent local move to be a function of life-cycle stage, housing tenure choice, earlier mobility experience, distance of the initial move, and local population gowth. The rationales for the first three of these independent variables are well established in the literature and will not be repeated here.

The model specifies that distance of the initial move has both a direct effect and an indirect effect on the probability of subsequent local mobility. If in-migrants are especially likely to rent, then they should have a higher rate of subsequent local mobility, since renters have much higher mobility rates than do owners. But the model also implies that, controlling for differences in initial tenure selection, in-migrants are more likely to move subsequently than are local movers.

Hypothesis 2. Both buyers and renters are expected to move again at a higher rate if the previous move was long distance than if it was local.

That is, whether or not the initial housing was perceived to be temporary, it often times turns out that way.

Hypotheses 1 and 2 are derived from the repeat mover theory. Our third hypothesis is derived from the place-effect theory.

Hypothesis 3. Families and individuals will have higher local mobility rates if they live in areas of rapid population growth, even after controlling for the effects of their own mobility histories.

To test our model we need a sample of households drawn from different SMSAs and providing detailed mobility histories. The Panel Study of Income Dynamics offers such a sample. The Panel Study is a longitudinal study consisting of annual interviews beginning in 1968 with a national probability sample of approximately 5,000 U.S. households (Survey Research Center, 1972). After application of a weighting factor to adjust for differential sampling rates and attrition, the sample is representative of the population of the U.S. exclusive of Alaska, Hawaii, and residents of most group quarters and institutions. Panel Study households that move are traced and remain in the sample.

The Panel Study definition of a move is based on the household head's response to the question, "Have you [head] moved since the spring of [previous year; approximately a one-year interval]?" For those reporting a move, the distance of the move is coded into three categories: local (intra-county), intra-state (inter-county within a state), and interstate. The subsample used in this analysis are households residing in an SMSA in 1970. The data are partitioned into two mobility intervals: (1) the 12 months preceeding the 1970 interview, referred to as the initial period, and (2) the 24 months preceeding the 1972 interview, referred to as the subsequent period. The household is the unit of analysis, and we will assume that all household members' mobility behaviors are the same as the head's, a tolerable simplification for our purposes. For more on the Panel Study and its application to the study of mobility sequences, see DaVanzo and Morrison (1981).

## FINDINGS

Several findings emerge from comparison of simple summary statistics from the sample (Table 9.1). The figures in the first row

Table 9.1: Mobility Behavior 1970-71, by Distance of Move in 1969

| | Distance of Move in Initial Period (1969) | | |
|---|---|---|---|
| | Local (intra-county; n = 650) | Long-Distance (inter-county; n = 174) | No Move (n = 2443) |
| Percentage of households that move in subsequent period (1970-71) | 57* | 73 | 20 |
| Percentage of households that move locally in subsequent period | 42 | 38 | 15 |
| Percentage of subsequent period non-migrants that move locally | 49* | 58 | 15 |

*Sample:* Households residing in an SMSA in 1970.

*Percentages for local and long-distance movers significantly different at .05 level.

show that long distance movers are more likely than are local movers to move again during the subsequent period. Furthermore, local and long distance movers both have substantially higher rates of subsequent mobility than do households that did not move during the initial period, a finding consistent with the previous literature on repeat mobility. The second row of Table 9.1 shows that local and long distance movers have comparable rates of subsequent local mobility, so the higher overall rate of subsequent mobility by long distance movers must be due to the long distance component.

The last row of Table 9.1 shows that when we restrict the sample to those who do not leave the local area during the subsequent period, initial long distance movers are almost 20 percent (58%/49% - 1.0) more likely to move again than are initial period local movers. This finding for the nation as a whole is consistent with the local area results reported by Roseman (1973) and Adams et al. (1973). But the simple statistics in the bottom row of Table 9.1 do not tell us whether the higher subsequent local mobility of long distance migrants is attributable to the distance of the previous move. To answer that question we turn now to estimation of our model.

The results from the estimation of the first equation (Table 9.2) support Hypothesis 1. The dependent variable is dichotomous--1 = move, O = stay. The dichotomous dependent variable is estimated by multiple classification analysis (MCA). MCA is a computationally convenient form of regression analysis, applicable to situations in which all of the independent variables are categorical. The unadjusted coefficients (gross effects) are simply the average deviations from the overall mean of the dependent variable, $\bar{y}$. The adjusted coefficients (net effects) are analogous to regression coefficients for dummy variables in a multiple regression, except that in MCA the deviations are measured from $\bar{y}$ rather than the mean value of y for the omitted dummy variable category. MCA's solution algorithm selects the set of adjusted coefficients that minimizes the sum of the squared errors in estimates of the dependent variables. For a detailed description of MCA, see Andrews et al. (1973). Statistical problems arise with ordinary-least-squares (OLS) regression analysis or with MCA when the dependent variable is dichotomous. Specifically, heteroscedasticity, biased estimates of standard errors, and estimated probabilities outside the 0-1 range are encountered. These problems have been discussed elsewhere (Goodman, 1976); that treatment can be summarized by stating that these problems leave the estimated coefficients unbiased, although they reduce the reliability of estimated standard errors and of any significant tests. The problems encountered with a dichotomous dependent variable are unlikely to affect the conclusions reached in this study.

Table 9.2 shows that interstate movers have a lower probability of buying their housing than do otherwise comparable movers who move a shorter distance. The adjusted deviations show that an interstate mover is only a little over half as likely as an intra-county

Table 9.2: Multiple Classification Analysis of Determinants of Tenure Choice By Movers (1 = buy; 0 = rent)

| Independent Variables | Unadjusted Deviations | Adjusted Deviations | Sample Size |
|---|---|---|---|
| Distance of Moves | | | |
| intra-county | .01 | .01 | 650 |
| intra-state | .08 | .05 | 71 |
| inter-state | -.11 | -.08 | 103 |
| Income | | | |
| 0 - $4,999 | -.13 | -.07 | 312 |
| $5,000 - $9,999 | -.02 | -.03 | 330 |
| $10,000 - $14,999 | .19 | .12 | 115 |
| $15,000+ | .38 | .25 | 67 |
| Life-Cycle Stage[a] | | | |
| YSNC | -.20 | -.13 | 139 |
| YMNC | -.01 | -.02 | 122 |
| YMC | .12 | .08 | 256 |
| OMC | .39 | .13 | 36 |
| OMNC | .32 | .19 | 30 |
| OSNC | -.06 | -.06 | 60 |
| YSP | -.09 | -.00 | 144 |
| OSP | -.16 | -.14 | 37 |
| Pre-Move Tenure | | | |
| own | .45 | .32 | 89 |
| rent or new household | -.05 | -.04 | 735 |

| | | |
|---|---|---|
| Mean value of dependent variable | .21 | |
| R-squared | .28 | |
| | **Eta** | **Beta** |
| Distance of move | .11 | .08* |
| Income | .37 | .24* |
| Life-cycle stage | .36 | .22* |
| Pre-move tenure | .38 | .27* |

*Sample:* Households residing in SMSAs in 1970 that moved in 1969.

[a]Key: Y = young (household head under 45); S = single (never married, divorced, separated, or widowed); M = married; NC = no children; C = children; O = old (household head 45+); P = parent.

*Beta significant at .95 level.

or intra-state mover to buy. The coefficients for the other three independent variables in Table 9.2 are as expected, based on the earlier discussion, and each is statistically significant. Since they are of interest here mainly as controls, they will not be discussed separately.

One of the reasons for expecting long distance movers to buy less frequently is that they may be uncertain whether they will be remaining in the area. The strength of this tendency for long distance migrants to migrate again is shown by the statistics in Table 9.1 and by previous studies of repeat migration. Assuming that there is some positive correlation between people's mobility plans and their subsequent behavior, we would expect that the effect of distance of move on tenure choice would be diminished by dropping from the sample those households that subsequently migrated out of the county. The presumption is that many of these households perceived that they would be migrating again and avoided homeownership for this reason.

The data support this expectation. Restricting the sample to those households that did not subsequently move out of the area reduced the differential between purchase rates of intra-county and interstate movers from nine percentage points to six, and distance of move no longer has a statistically significant effect on tenure choice. (Extended tables which include these results are available from the author.)

The second equation of the model tests Hypothesis 2--that distance of move has an independent effect on subsequent local mobility, in addition to the expected indirect effect operating through tenure choice. The results of this estimation (Table 9.3) offer some support for this position. Controlling for tenure choice, life-cycle stage, and earlier mobility experience, we see that interstate and intra-state movers in the sample are more likely to move again, within the area, than are households whose previous move was local. So distance of initial move has a direct effect on subsequent local mobility propensity, as well as an indirect effect operating through tenure choice. The statistically significant coefficient on the distance-of-initial-move variable implies that housing that turns out to be temporary is not always initially perceived that way, since home purchase is unlikely for someone expecting to move again in the next year or two.

The second equation also supports Hypothesis 3--local population growth promotes local mobility. Table 9.3 shows that households in growing areas are seven to eight percentage points more likely to move locally over the two-year period than are otherwise comparable households in declining counties.

## CONCLUSIONS

We have found empirical support for both the repeat mover theory and the place-effect theory. Both explanations contribute to

Table 9.3: Multiple Classification Analysis of Determinants of Subsequent Local (Intra-County) Mobility (1 = local move; 0 = stay)

| Independent Variables | Unadjusted Deviations | Adjusted Deviations | Sample Size |
|---|---|---|---|
| Distance of Initial Move | | | |
| intra-county | .21 | .09 | 587 |
| intra-state | .29 | .15 | 51 |
| inter-state | .33 | .14 | 62 |
| non-mover | -.07 | -.03 | 2330 |
| Tenure Choice | | | |
| own | -.19 | -.16 | 1446 |
| rent | .17 | .14 | 1584 |
| Life-Cycle Stage[a] | | | |
| YSNC | .27 | .08 | 192 |
| YMNC | .20 | .11 | 171 |
| YMC | .02 | .03 | 916 |
| OMC | -.18 | -.07 | 408 |
| OMNC | -.15 | -.05 | 376 |
| OSNC | -.09 | -.10 | 340 |
| YSP | .14 | .05 | 434 |
| OSP | -.02 | -.04 | 193 |
| Year of Most Recent Move Prior to 1969 | | | |
| 1968 | .20 | .10 | 592 |
| 1965-67 | .05 | -.00 | 970 |
| before 1965 | -.11 | -.04 | 1468 |
| County Population Growth, 1960-1970 | | | |
| <0% | .00 | -.06 | 631 |
| 0-14% | -.00 | .01 | 1130 |
| 15%+ | -.00 | .02 | 1269 |

| | | | |
|---|---|---|---|
| Mean value of dependent variable | | .28 | |
| R-squared | | .23 | |
| | Eta | | Beta |
| Distance of move | .27 | | .12* |
| Tenure choice | .40 | | .33* |
| Life-cycle stage | .30 | | .14* |
| Year of most recent move | .27 | | .11* |
| County population growth | .00 | | .07* |

*Sample:* Households residing in SMSAs in 1970 that did not leave the SMSA before 1972.

[a]For key, see Table 9.2.
*Beta significant at .95 level.

the correlation between an area's in-migration rate and its local mobility rate.

The distance of their move as well as the mere act of moving contribute to migrants' high rate of subsequent local mobility. Many migrants initially select temporary accommodations and soon thereafter move to more permanent accommodations within the metropolitan area.

Independent of their own mobility history, households in high-growth counties have above-average local mobility rates. High rates of in-migration can change communities and neighborhoods in ways that induce current residents to move out. The high rates of housing turnover accompanying population growth expose current residents to a greater number of alternative housing opportunities, increasing the chances that one will be chosen over the current house or apartment.

Our micro-estimates of repeat mover and place effects are roughly consistent with the magnitude of the macro-correlation between migration rates and local mobility rates shown in Figure 9.1. However, differences in calibration prohibit precise comparisons. Yet in Figure 9.1 only about 60 percent of the variation in local mobility rates can be explained by the long distance rates. Clearly, housing market, labor market, and population differences unrelated to in-migration also affect local mobility rates.

Substantive and methodological implications emerge from our findings. First, we have shown that the relocation costs for many migrants exceed the cost of the initial move into the local area. Our results indicate that perhaps 10 percent of the subsequent local moves by interstate movers are either directly or indirectly attributable to their having acquired temporary accommodations upon moving into the area. To the extent that these follow-on costs enter into migration decision-making, they should be incorporated into existing micro-level migration models.

A methodological implication is that previous models of home purchase decisions may be subject to a specification bias. We have shown that distance of move has an effect on tenure choice. Since distance of move is positively correlated with household income, the independent effect of income on movers' home purchase decisions will be understated in models in which distance of move is not included as an independent variable. This bias is especially important in the application of models of residential mobility and housing choice to analysis of public policy options such as housing allowances.

From the viewpoint of municipal and neighborhood officials interested in promoting population stability, our results imply that local movers are preferable to long distance migrants as buyers or renters of housing coming on the market within their jurisdiction. Migrants are more likely to soon leave the area than are local movers, and they can also be expected to move locally (but perhaps out of the neighborhood or municipality) at a higher rate than otherwise comparable local movers.

Linking local mobility rates

Perhaps the most important implication of our analysis is that future changes in rates of local mobility in the U.S. will depend on future long distance migration rates. Long and Hansen (1977) have argued that several long-run changes in the nation's economy and population will lead to future declines in long distance migration. If and when this decline begins, the nation's local mobility rate can be expected to decline as well.

## REFERENCES


Adams, J. S. et al. (1973) "Commentary: intra-urban migration", Annals of the Association of American Geographers, 63, 152-155.

Andrews, F. et al. (1973) Multiple Classification Analysis. Ann Arbor: University of Michigan, Institute for Social Research.

Da Vanzo, J. (1978) Repeat Migration In The United States: Who Moves Back And Who Moves On? Santa Monica, California: The Rand Corporation, P 5961.

----- and P. Morrison (1981) "Return and other sequences of migration in the United States", Demography, 18, 85-102.

Frey, W. H. (1979) "Central city white flight: racial and non-racial causes", American Sociological Review, 44, 425-448.

Goodman, J. L. Jr. (1976) "Is ordinary least squares estimation with a dichotomous dependent variable really that bad?", Washington, D.C.: The Urban Institute, Working Paper.

----- (1979) "Reasons for moves out of and into large cities", Journal of the American Planning Association, 45, 407-416.

Long, L. H. and K. Hansen (1977) "Migration trends in the United States since 1935", paper presented at the International Union for the Scientific Study of Population, Mexico City.

----- and C. G. Boertlein (1977) "Urban residential mobility in comparative perspective", Washington, D.C.: U.S. Bureau of the Census, mimeo.

MacMillan, J. (1980) Mobility in the Housing Allowance Demand Experiment. Cambridge, Mass.: Abt Associates.

McCarthy, K. (1976) "The household life cycle and housing choices", Papers, Regional Science Association, 37, 55-80.

Miller, A. R. (1967) "The migration of employed persons to and from metropolitan areas of the United States", Journal of the American Statistical Association, 62, 1418-1432.

Morrison, P. A. and D. A. Relles (1975) Recent research insights into local migration flows. Santa Monica, California: The Rand Corporation, P 5379.

Roseman, C. C. (1971) "Migration as a spatial and temporal process", Annals of the Association of American Geographers, 61, 589-598.

----- (1973) "Comment in reply", Annals of the Association of American Geographers, 63, 155-156.

Rossi, P. H. (1955) Why Families Move. Glencoe: Free Press.

Speare, A., S. Goldstein, and W. H. Frey (1975) Residential Mobility, Migration, and Metropolitan Change. Cambridge, Mass.: Ballinger.

Stone, L. O. (1971) "On the correlation between metropolitan area in- and out-migration by occupation", Journal of the American Statistical Association, 66, 693-701.

Survey Research Center (1972) A Panel Study of Income Dynamics. Ann Arbor: University of Michigan, Institute for Social Research.

Weicher, J. C. (1977) "The affordability of new homes", American Real Estate and Urban Economics Journal, 5, 209-226.

# 10

# SEARCH BEHAVIOR AND PUBLIC POLICY: THE CONFLICT BETWEEN SUPPLY AND DEMAND PERSPECTIVES

**Eric G. Moore**

From the perspective of the planner or policy-maker in the housing arena, particularly in individual cities, recent research on residential search poses a major problem of interpretation. The decision-maker's task is to identify and implement strategies for dealing with housing problems which are often defined in terms of the cost or quality of housing services consumed by small and highly specific populations within his jurisdiction. The available tools for dealing with these problems are limited and tend to be concentrated on the supply-side of the market equation (construction incentives, price control, public housing, zoning, code enforcement, anti-discrimination codes, fair lending practices); only recently has the growth in interest in housing allowances introduced a major demand-side option.

In deciding on an appropriate action, one should understand the relationship between the available programmatic tools and their ability to induce changes in cost, quality and other measures of housing consumption. Unfortunately, the overwhelming emphasis on consumer behavior in recent studies produces few insights in this regard such that the links between research on residential search and problem solving in the housing sphere are often tenuous. In this chapter, I argue that if the policy implications are to be understood, a more broad-based approach to search must be adopted which regards the acquisition and utilization of information about housing alternatives as the outcome of the interaction between management and policy decisions and consumer actions in an imperfect market.

## RECENT STUDIES OF SEARCH

The recent upsurge in research on the relation between search behavior and residential mobility is dominated by a standard paradigm. This paradigm has been adopted (with only minor variations) in research associated with the large-scale housing allowance experiments funded by the U.S. Department of Housing and Urban Development; it is these experiments which have been a major source for papers on search behavior in the last three years (e.g.

 DOI: 10.4324/9781003182085-12

Weinberg et al., 1977; Cronin, 1979, 1980; McCarthy, 1979). A central argument in this work is that a stimulus to search does not necessarily lead to a residential relocation and, thus, the decision to search should be separated analytically from the decision to move. The general view of this process has the following components (Figure 10.1).

(1) The decision to search is represented as a function of the difference between current consumption and preferred (or equilibrium) consumption modified by perceived frictional costs. Here, much of the debate concerns the measurement of preferred or equilibrium consumption; the direct approach of Maclennan and Wood to characterizing desired consumption of households can be contrasted with the attempts to define surrogate populations at equilibrium in Hanushek and Quigley (1978) and Weinberg, Friedman and Mayo (1979).

(2) The search process is characterized in terms of the strategies employed in collecting and utilizing information about potential alternatives and of the costs incurred in these efforts. Attempts are made to identify sources of information used in detecting vacancies, their relative effectiveness in uncovering acceptable alternatives together with measures of the length, intensity and spatial domain of the search activities. Using these measures as independent variables, various authors seek to model their variation as functions of both the socio-demographic attributes of the searchers, their resources and the role of institutional variables such as availability of credit, help from government agencies and discrimination by owners and managers (see the Introduction to this volume and Clark (1981) for a survey of this work).

(3) The outcome of the search process is characterized at two levels: the decision to move or to stay in the existing dwelling and the attributes of the new consumption bundle (the latter include neighborhood and access characteristics as well as dwelling condition).

Most recent research effort has been devoted to differentiating the structure of the search process for various population sub-groups defined along racial, income or program eligibility lines. Only a limited amount of work has been undertaken on the relation between search and consumption outcomes; in particular the role of specific search parameters, such as length or intensity of search in affecting these outcomes seems to be minimal.

In adjudicating the effectiveness of particular public interventions, the impact on outcomes is a critical, though not the only, issue. It is also necessary to recognize that perceived "equality of treatment" is politically important (Grigsby and Rosenberg, 1975) a situation which generates some concern for easing problems of

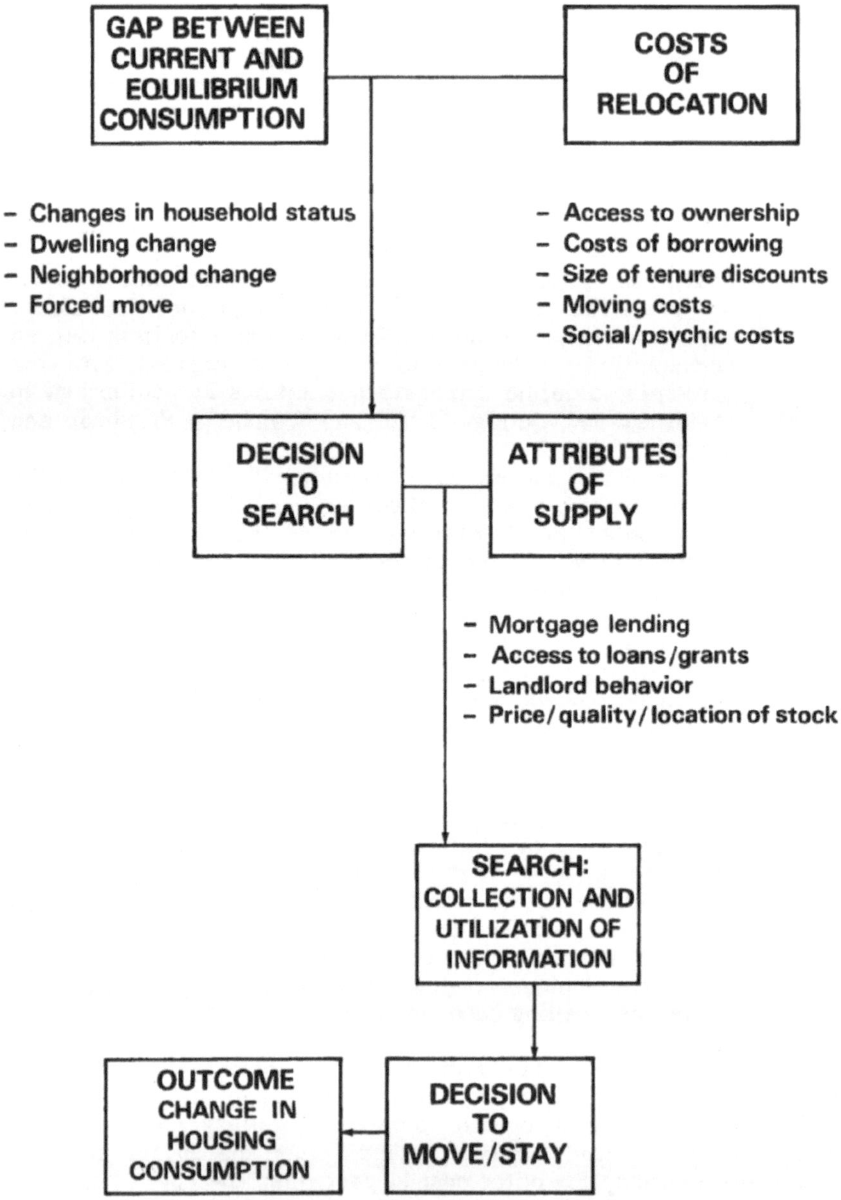


Figure 10.1. A neoclassical framework for the analysis of movement decisions.

search per se; however, the prime need is to address consumption issues. It is possible that substantial out-of-pocket costs might significantly affect the resources available for purchase of housing or other goods, but the most central concerns are to assess the efficacy of residential adjustments in providing benefits to the occupant at the new location and to understand the constraints which prevent households from entering the market in the first place.

In organizing the subsequent discussion, two issues dominate: the first, and shorter section, considers the nature of outcomes which are of concern in the public policy domain. It is these outcomes which must be linked to policy tools. The second section focuses on the organization of information flows within the housing market. It is important to remember that there are many actors other than the household in this market-- landlords, property owners and managers, real estate agents, financial institutions, public agencies--whose goals are not necessarily identical to those of households (Bassett and Short, 1980). Management of information can serve their interests and have considerable effects on observed search behavior and its outcomes. A better understanding of this interaction between supply and demand in the use of information is a necessary precursor to better decision support for planning and policy making in the public sector.

## THE NATURE OF OUTCOMES

Intrinsic to many discussions of mobility behavior, ranging from the seminal work of Rossi (1955) to the most recent models of household disequilibrium (Hanushek and Quigley, 1978) is the notion that mobility is the primary mechanism for housing adjustment, at least in the rental market. This basic belief underlies a great deal of federal housing policy in the last three decades which seeks to facilitate this adjustment process through mobility (Hartling, 1980). Yet, apart from a tautologous argument which can all too easily arise within a utility-maximizing framework, how do we tell if consumption is enhanced by relocation? Housing is a complex, multi-faceted concept embracing not only relatively easily measured items of size, cost, condition, and location but also less tangible notions of privacy, status, satisfaction and future security. If we accept this characterization of housing, it is clear that its unidimensional treatment in terms of expenditure embodying, as it does, Olsen's (1967) concept of a unit of housing service in a perfectly competitive market, is an inadequate policy tool.

At this point in time, we can do no more than suggest elements of outcome evaluation which should be considered, since no coherent framework outside of neoclassical micro-economics has yet been developed. Furthermore, it should be recognized that any attempt to measure outcomes must surely imply a socio-political stance. Use of a single expenditure measure implies a fundamental commitment to market mechanisms as a solution to housing problems: broader-based measures incorporating indicators of size, quality and access (e.g. Goodman, 1976) still link evaluation to a materialist view of the importance of consumption. Criteria such as those used by Grigsby and Rosenberg (1975), which emphasize the importance of comparison of consumption variables over different socio-economic and demographic groups, focus on the political issue of "equality of treatment." Analyses which emphasize the increasing concentration of those with serious housing problems within particular classes or class fractions (Dear, 1980) tend to adopt evaluative criteria which stress more radical redistributional needs in the larger society.

Particularly from the point of view of the public decision maker, rather than the academic, the greatest need appears to be that of characterizing the trade-offs which are made between affordability and attributes of the consumption bundle for different politically visible population sub-groups (a combination of the Goodman and Grigsby criteria). The competitive market view continues to pose real problems of interpretation when set against analyses of data from admittedly imperfect markets, while Marxist-oriented redistributive proposals immediately confront problems of implementation within the existing institutional structure.

In developing outcome measures, therefore, the emphasis initially is on the characterization of an array of variables including size, quality, access, neighborhood composition and associated cost and cost-to-income indices for those making relocation decisions in individual markets. Efforts then need to be made to categorize outcomes, possibly in terms of their level of ambiguity (e.g. unambiguously superior after relocation, when improvement is made on each variable; ambiguous outcomes with mixtures of increasing and decreasing measures; and unambiguously inferior when measures on all variables get worse). The task is then one of seeking to characterize the personal and institutional circumstances surrounding those situations in which relocation (a) fails to lead to improvement in consumption above those levels defined as "problems," or (b) produces unambiguously inferior results.

At a crude level, available data from the Community Development Block Grant Program evaluation (Moore, 1980) emphasize the general ambiguity of relocation outcomes across nine cities. This finding reinforces earlier statements about the Housing Allowance Experiments (e.g. Rossi, 1979). Table 10.1 indicates that in 7 of the 9 CDBG sample cities, movers are experiencing noticeable increases in the amount spent on rental housing and that trade-offs with respect to other aspects of housing are made. In Denver, for example, the considerable inflationary pressure on rents leads to a difficult situation for movers reflected not only in the change in amount spent, but a slight decrease in quality and low levels of expressed satisfaction with dwellings after relocation.

These outcomes of search are influenced by the type and quality of information available to those looking for new dwellings in the housing market. For the policy analyst, the question is whether a

Table 10.1: Outcomes of Residential Moves for Renters: CDSE Study, 1979

| City | Tenure Discounts $ | Percent Increase in Rent After Move | Percent Increase in Dwelling Quality for Movers | Mover Satisfaction | | Non-Mover Satisfaction | |
|---|---|---|---|---|---|---|---|
| | | | | Dwelling % | N'hood % | Dwelling % | N'hood % |
| Birmingham | 58[a] | 39 | 47[b] | 59[c] | 53 | 43 | 60 |
| Corpus Christi | -12 | 0 | 14 | 30 | 55 | 50 | 68 |
| Denver | 75 | 47 | -3 | 26 | 68 | 59 | 60 |
| Memphis | 12 | -8 | -6 | 28 | 48 | 44 | 55 |
| New Haven | 10 | 55 | -4 | 40 | 46 | 52 | 35 |
| Pittsburgh | 12 | 18 | 23 | 60 | 50 | 58 | 53 |
| St. Paul | 44 | 29 | -3 | 59 | 71 | 48 | 76 |
| San Francisco | 18 | 16 | -1 | 60 | 54 | 51 | 49 |
| Wichita | 21 | 46 | 8 | 46 | 44 | 43 | 50 |

a. Tenure discounts are increases in rent of a standard unit for movers compared to non-movers over the previous 12 months (see Moore, 1980).

b. The quality measure is a synthetic number representing average improvement over a range of physical attributes.

c. For satisfaction variables the number records the proportions who say they are "very satisfied" with their dwelling or neighborhood.

better understanding of search behavior can lead to identifiable - strategies for improving information access such that households experiencing major consumption difficulties are helped. However, in order to do this, we need to know more about the institutional arrangements which guide the flow of information about housing opportunities.

## THE ORGANIZATION OF INFORMATION FLOWS IN THE MARKET

In trying to understand how particular outcomes are achieved in given market, it is not unreasonable to suggest that the availability and utilization of information about alternatives are influential. The problem is in identifying the way in which they are influential. In the standard demand-oriented paradigm used in many analyses of search behavior, the problem is treated as one of consumer access to information which, implicitly, just exists in the prospective search environment. Attributes of the household search process are then considered to be functions of the characteristics of the households and their resources. Such a view raises difficult problems of interpretation.

For illustrative purposes, consider the regression equation in McCarthy's study (1979, 27) which links rent discounts to parameters of search behavior (Table 10.2). The attempt to define such a link is laudable but the meaning of the parameter values in the estimated equations is far from clear. The dichotomous variable (use of low intensity search among friends) is associated with the highest rent discount of about 6 percent. Is this meant to imply that, from a consumer perspective, one is more likely to uncover a bargain if you use your friends or does it say more about the way in which certain transactions are handled by the suppliers of rental units? Perhaps the observed behavior reflects a situation in which landlords place high value on "good tenants" who will, in aggregate, possess low turnover rates and will help to maintain the condition of the property. Under such circumstances, landlords will both be willing to offer discounts for "good tenants" and would be more likely to try to find them through friends or existing "good tenants" than through less personal channels.

The latter interpretation leads to a recognition of a different class of problem. A group of households labelled as "poor tenants," for whatever reason, are likely to be present in any pool of renters. Use of inter-personal contacts to acquire information is unlikely to have much effect as the major problem lies in responses of those on the supply side. A more promising approach might follow the suggestion of Grigsby and Rosenberg (1975) that emphasis should be placed on the development of appropriate organizations and of management skills necessary to cope with "problem tenants".

The main point behind the above discussion is that observations on how individuals search for housing are to be explained as much by actions of those supplying information as of those consuming it. If

Table 10.2: Rent Discount Equations: Renter Households in Brown County, Wisconsin

| Variable | Coefficient | Value of t |
|---|---|---|
| Dependent | | |
| Monthly rent discount (%) | -- | -- |
| Independent | | |
| Constant | 3.05 | .55 |
| No active search (yes = 1) | -.10 | 1.62 |
| Units examined (ln) | .86 | 1.25 |
| Search length (ln) | -.27 | .78 |
| Sources used | .11 | .46 |
| Problems encountered (ln) | 2.88 | 2.31[a] |
| Low intensity -- friends (yes = 1) | -5.96 | 3.30[a] |
| High intensity -- no problems (yes = 1) | -1.63 | .37 |
| Background | | |
| Local moves | -.38 | .71 |
| Length of stay (ln) | -.52 | 1.11 |
| Head's years of schooling | .61 | 2.40[a] |
| Single male head (yes = 1) | 1.00 | .49 |
| Single female head (yes = 1) | .45 | .22 |
| Single head -- children (yes = 1) | 1.16 | .41 |
| Single person household (yes = 1) | -5.34 | 2.76[a] |
| Age of Household Head | | |
| <21 (yes = 1) | -2.59 | .94 |
| 21-29 (yes = 1) | -6.15 | 2.98[a] |
| 30-39 (yes = 1) | -.80 | .34 |
| 40-69 (yes = 1) | -2.73 | .79 |
| 70+ (yes = 1) | 3.61 | .81 |
| Number of children | .75 | 1.32 |
| Income eligible (yes = 1) | -4.71 | 3.17[a] |
| Near eligible (yes = 1) | -1.15 | .80 |
| Income Sources (%) | | |
| Welfare | .05 | 1.28 |
| Pensions & social security | -.03 | .61 |
| Earnings | -.03 | .86 |
| $R^2$ | .113 | |
| F | 4.64 | |

*Source:* Adapted from: McCarthy, 1979, p. 27.

*Note:* Regression analysis was performed on records of 933 renter households paying full market rent and moving locally in the 5 years preceding the survey.

[a]Coefficient is significantly different from zero at the .05 level.

we consider, for example, the observation that significantly more nonminority than minority households use newspapers while minority searchers make greater use of vacancy signs (Cronin, 1979), we might ask to what extent these differences reflect the advertising strategies of different groups of landlords. Certainly in large cities, daily newspapers tend to focus on the middle to upper price ranges while weekly neighborhood papers cover more modest units and vacancy signs and cards on bulletin boards address the low end of the market. If this scenario is correct, not only would this imply a variation in sources used by those with different incomes but it would also imply differences in extent and intensity of search; at the low end of the market, information is much more poorly co-ordinated and its spatial fragmentation is likely to lead to more localized patterns of search. Further, the same amount of time and effort applied in different segments of the market is likely to produce differential levels of success measured in terms of attainment of desired consumption.

Given this general view of the importance of supply side activities, we now consider more specific impacts of activities on the part of institutions in both the public and private sectors.

## Consequences of Government Activities

Government involvement in housing in different countries occurs in many forms ranging from massive amounts of direct intervention in both construction and management as in the Netherlands and Britain to very limited activity as in the United States. In the former case, the entire nature of the search process can be transformed. In Britain, where over 30 percent of housing is in the public sector and, in specific urban housing market over 80 percent of new construction is in this sector, relocation into the public sector is by allocation and within the public sector is by mutual exchange. The allocation involves no search at all (and only a limited ability to reject offers) while, in the transfer process, search is transformed by the requirement that you must find someone who is willing to trade his or her house for yours (Bird, 1976). In fact, the nature of this requirement is such that it probably eliminates much of the strategy of search described in the North American literature; it becomes more like buying tickets in a neighborhood lottery in which you win if you can find the person with the other half of your ticket.

The effect on search is similar in the Netherlands where permits for occupancy are required by both landlords and tenants in the price controlled sector which occupies 60-80 percent of the market. Tenants are given permits to occupy specific types of units based on the household's characteristics (size, age, marital status) while landlords have permits to rent to particular types of households, given the unit attributes (size, structure type, the floor on which it is located). Search then involves finding a vacant unit whose landlord permit matches your tenant permit. In a system in which mobility is very low and new opportunities scarce, search is made more difficult.

Strict rent control is also a major feature of the Dutch market and, given the growing incidence of rent control in both the U.S. and Canada, it is appropriate to question its impact on search. The evidence is quite persuasive that rent control leads to substantial reduction in overall mobility (e.g. Clark and Heskin, 1981). The number of opportunities available to the searcher are therefore more limited and the likelihood that a given stimulus will initiate search is also reduced. For those who do search, additional information on vacancies becomes desirable, particularly in regard to the application of local rent control legislation to a specific unit (for example, to identify the level of rent increase permitted for that unit upon change of tenure). More skillful or knowledgeable searchers might utilize a much wider range of legal or governmental sources in search than are included in the standard list of information sources in the current search literature.

At the extreme case where rent control exists in a highly inflationary market such as Los Angeles, California, the market may become so tight (with vacancy rates of less than 1 percent; Clark et al. (1980)) that queues begin to form in the private sector. For desirable apartments in attractive locations it becomes a sellers market and landlords or managers can choose from a waiting list of "desirable" tenants. As in the case of queues forming for entry into the public housing sector (Bird, 1976), search is primarily a function of waiting in line.

It is clear that many other Government activities affect specific types of search. Government programs which provide loans or grants for acquisition or improvement of dwellings contain eligibility rules and provide specific information relative to search and choice (such programs include assisted homeownership programs, rehabilitation loan and grant programs and rent-subsidy payments all of which exist in some form in both the U.S. and Canada). Although these programs, together with the more general classes of Government intervention discussed above, are of direct concern to public policy, the bulk of the search literature has avoided the issue of their effect on information availability to different sub-groups in the housing selection process. Most work has focussed on private sector relocations; however, it soon becomes apparent that when we consider supply side issues in this sector, similar problems arise in assessing the impacts of various institutional activities on observed consumer behavior.

## Information Flows in the Private Sector

A wide variety of actors are involved in the supply of private sector housing: developers, real estate agents, mortgage companies, banks, landlords and property owners and managers. While all seek to enhance their profits in one way or another, they may take quite different perspectives on how this is to be achieved. For some, short term gains through single sales are important; for others, building professional confidence and a steady future stream of clients is regarded as the proper strategy. Banks and mortgage companies seek to reduce risk by careful screening both of potential borrowers and of the prospects for future change in value of the mortgaged property. Some landlords seek higher rents and higher turnover, while others try to reduce the losses due to frequent vacancies by offering discounts to more stable tenants. For some property managers the goal might be to promote high rents solely for the purpose of raising the assessed value of the property such that more money can be borrowed for other investment purposes (Bassett and Short, 1980). These different goals lead to a plethora of management strategies which must surely affect the search behavior of consumers and certainly influence the nature of outcomes.

Each supplier chooses a strategy for reaching the group which he would identify as his "clients". This strategy is subject to economic criteria in the sense that the costs of a given procedure for finding buyers or tenants must be commensurate with the returns. In the rental market, for example, agency listing involves payment of a commission which is only worthwhile if the tenants pay a reasonable amount of rent for an extended period: short lets at low rentals would not justify the costs of using an agency. Similarly, real estate agents offering a sophisticated service in the high priced end of the market are likely to place lower bounds on the prices of units listed.

The information available in the various channels identified in the search literature (newspapers, real estate agents, bulletin boards, friends and relatives, vacancy signs) tends to be more strongly associated with some segments of the market than with others. Thus, it is reasonable to suggest that the choice of information channel is not independent of the segment of the market in which the searcher is looking, a factor which has received very little attention in the literature.

On a priori grounds, I would also speculate that strategies used by given suppliers are likely to shift as external market conditions change. In a sellers market, for example, multiple listing is likely to be less effective sinces sales will accrue to specific realtors prior to notification of listing being fed into the broader network. In a seller's market, more units are likely to change hands through word of mouth and the consumer takes greater risks in not accepting attractive opportunities as soon as they are found. In the rental sector, low vacancy rates mean landlords can target advertising to more specific groups and, when the market becomes exceedingly tight, the formation of queues puts landlords in a dominant role and they can be highly discriminatory with respect to those with poor tenancy records.

Apart from the management of information flows, the activities of agents intermediary between seller and buyer have strong influences on the structure of search. The role of finance and mortgage companies in controlling access to opportunities for given income and socio-demographic groups has received considerable attention in studies by Harvey (1974), Boddy (1976) and Stone (1978). More recently, a number of writers have focussed on the role of real estate agents in structuring household decisions (Palm, 1976; Bordessa, 1978; Spector, 1979). Two activities of agents are particularly important in their effect on search.

(1) The explicit strategy in presenting houses to prospective buyers: the number shown is limited and the order presented maximizes the chance of a unit being accepted after only four or five houses are visited (see Smith's (1979) discussion of the "hammer and anvil" and Shenkel's (1979) treatment of the Burke method). Clearly, such strategies tend to produce very strong structure in reported search behavior.

(2) The attempt to organize the market at the neighborhood level: specific areas become identified with particular stages in family development and social status. Thus the possibilities of renewed business through "trading-up" are promoted (Spector, 1979). Again the extent to which these activities are successful will impose a strong structure on the spatial nature of search.

## THE IMPLICATIONS FOR PUBLIC POLICY

The preceding discussion brings us to a major issue which needs to be confronted: what type of explanation of search and housing adjustment behavior is likely to contribute to design of intervention strategies in the public sector. As has been discussed in an earlier paper (Moore and Clark, 1980), public policy is designed to meet a variety of social and political goals, including that of improving the housing consumption of low and moderate income households. Problems are defined in terms of the levels and distribution of housing consumption and available intervention tools are largely indirect and concentrated on the supply-side.

Conventional demand-oriented analyses of search behavior assume the supply of housing to be given. The nature of the ensuing explanation seeks to account for the way in which alternatives are uncovered and choices made between these alternatives; by its very nature, this exercise has nothing to say about situations in which households are allocated to dwellings nor about how supply conditions arise which constrain the search process and its outcomes. The thrust of my argument has been that the latter circumstances must be contained within the explanation of housing market processes rather than as a set of exogenous conditions for explanatory models of individual behavior. The reason for this is that problems of consumption arise out of the complex interaction between supply and demand at the local level and that reasonable solution strategies are unlikely to arise from conceptualizations which focus on only one side of the market equation. This position is re-inforced by the observations that the tools of intervention are predominantly supply-side activities and that recent demand-side interventions have had remarkably little impact on consumption adjustments (Hanushek and Quigley, 1979).

One of the major consequences of the various supply-side activities in managing information flows is that consumer-oriented analyses of search are difficult to interpret for policy purposes. They certainly provide descriptive statements characterizing differences in search experience of different population sub-groups and, as such, can provide important clues to the problems being encountered. However, the absence of a supply-side argument makes it difficult to define program strategies which are likely to ameliorate consumption problems to any significant degree.

If it is true that supply-side activities exert a major influence on the availability of information, then in many cases we can regard information as an allocation device used by suppliers to ensure that they provide housing services to the desired consumers. Treating search as a consumer problem leads to recommendations that the flow of information be improved (Weinberg et al., 1977); yet, is this a logical solution to the general consumption problem or is it only relevant to the modelled situation in which supply is assumed to respond to demand. If the goals of supply-side actors are to try to capture particular segments of the consumer population, increasing information is only likely to produce limited inter-personal shifts in outcomes and not substantial increases in the overall consumption of housing. In order to accomplish the latter, the major emphasis must be on changing the structure of opportunities being provided to different sub-groups; to develop strategies for accomplishing this, public decision-makers must have a better understanding of the nature of supplier behavior. This requires that supply-side activities be part of the explanation and not exogenous to it.

In fact, if public information is to be provided to consumers, it would seem more appropriate to follow Goldberg and Horwood's (1978) suggestion that information should be provided about the actors in the transaction process (real estate agents, lawyers, lenders, appraisers) rather than opportunities per se. Costs are often highly variable within local markets and significant saving can be attained by better knowledge of conveyancing procedures.

This discussion is very much in the spirit of Scriven's (1969) argument that the nature of an explanation is a function of the intent of the question. When the primary focus is on characterizing cognitive processes in household search, the existing literature clearly provides an excellent source of explanatory material. When the intent of the research endeavour is to develop intervention strategies for improving housing consumption with respect to a given criterion, then the resulting explanations must contain those elements which are relevant to the decision-maker's feasible actions. In the present case, this requires that explanations of search behavior place much greater emphasis on the behavior of supply-side actors than has been the case up to now.

## REFERENCES


Bassett, K. and J. R. Short (1980) Housing and Residential Structure: Alternative Approaches. London, Routledge and Kegan Paul.

Bird, H. (1976) "Residential mobility and preference patterns in the public sector of the housing market", Transactions of the Institute of British Geographers, New Series, 1, 20-33.

Boddy, M. (1976) "The structure of mortgage finance: building societies and the British social formation", Transactions of the Institute of British Geographers, New Series, 1, 58-71.

Bordessa, R. (1979) "Real estate salesmen and residential relocation decisions", The Canadian Geographer, 22, 334-8.

Clark, W. A. V. (1981) "On modelling search behavior", in D.A Griffith and R. MacKinnon (eds.) Dynamic Spatial Models. The Netherlands: Sijthoff and Noordhoff.

----- and A. Heskin (1982) "The impact of rent control on tenure discounts and residential mobility", Land Economics, 58 (forthcoming).

-----, A. Heskin and R. Manuel (1980) Rental Housing in the City of Los Angeles. Los Angeles, California: University of California, Institute for Social Science Research.

Cronin, F. J. (1979) "An economic analysis of intra-urban search and mobility using alternative benefit measures", Washington, D.C.: The Urban Institute, Working Paper.

----- (1980) "Racial differences in the search for housing", Washington, D.C.: The Urban Institute, Paper 1510-4.

Dear, M. J. (1980) "The public city", in W. A. V. Clark and E. G. Moore (eds) Residential Mobility and Public Policy. Beverly Hills, California: Sage.

Goldberg, M. A. and P. Horwood (1978) Housing Transactions Costs. Vancouver: Michael A. Goldberg, Urban and Economic Consultants.

Goodman, J. (1976) "Housing consumption disequilibrium and local residential mobility", Environment and Planning A, 8, 855-74.

Grigsby, W. and L. Rosenberg (1975) Urban Housing Policy. Center for Urban and Policy Research, Rutgers University, New York: APS Publications.

Hanushek, E. A. and J. M. Quigley (1978) "An explicit model of intra-metropolitan mobility", Land Economics, 54, 411-29.

----- (1979) Complex housing subsidies and complex household behavior: consumption aspects of housing allowances, New Haven: Yale University, Institution for Social and Policy Studies, Working Paper 825.

Hartling, J. E. (1980) "The public policy environment: mobility researchers' responsibilities" in W. A. V. Clark and E. G. Moore (eds.) Residential Mobility and Public Policy. Beverly Hills, California: Sage.

Harvey, D. (1974) "Class-monopoly rent, finance capital and the urban revolution", Regional Studies, 8, 239-55.

Maclennan, D. and G. A. Wood (1980) "Housing market search models: some empirical tests", paper presented at the Seminar on Models of Search Behavior, U.C.L.A. Los Angeles, California.
McCarthy, K. F. (1979) Housing Search and Mobility. Santa Monica: Rand Corporation. R-2451 HUD.
Moore, E. G. (1980) Community Development Strategies Evaluation: Household Relocation in the Sample Cities, Philadelphia: University of Pennsylvania, Department of Regional Science.
----- and W. A. V. Clark (1980) "The policy context for mobility research" in W. A. V. Clark and E. G. Moore (eds.) Mobility and Public Policy. Beverly Hills, California: Sage.
Olsen, E. O. (1967) "A competitive theory of the housing market", American Economic Review, 57, 612-621.
Palm, R. (1976) "The role of real estate agents as information mediators in two American cities", Geografiska Annaler, 5B, 28-41.
Rossi, P. H. (1955) Why Families Move, Glencoe: Free Press.
----- (1979) "Housing allowances and residential mobility", paper presented at the Conference on the Experimental Housing Assistance Program, The Brookings Institution, Washington, D.C.
Scriven, M. (1969) "The covering law position: a critique and an alternative analysis", in L. I. Krimmerman (ed) The Nature and Scope of Social Science: A Critical Anthology. New York: Appleton Century Crofts.
Shenkel, W. M. (1979) The Real Estate Professional. Homewood, Illinois: Dow-Jones Irwin.
Smith, T. R. (1979) "Agent influence in the housing choice process", paper presented at the Annual Meetings of the Association of American Geographers, Philadelphia.
Spector, A. N. (1979) "The real estate agent, the search for housing and the development of housing markets", paper presented at the Annual Meetings of the Canadian Association of Geographers, Victoria, B.C.
Stone, M. E. (1978) "Housing, mortgage lending and the contradictions of capitalism", in W. K. Tabb and L. Sawyers (eds) Marxism and the Metropolis. New York: Oxford University Press.
Weinberg, D., R. Atkinson, A. Vidal, J. Wallace and G. Weisbrod (1977) Locational Choice, Part 1: Search and Mobility in the Housing Allowance Demand Experiment. Cambridge: Abt Associates.
----- and J. Friedman and S. Mayo (1979) "A disequilibrium model of housing search and residential mobility". Cambridge, Mass.: Abt Associates.

# THE CONTRIBUTORS

W.A.V. CLARK is Professor of Geography at the University of California, Los Angeles. His research interests are focused on intraurban migration and neighborhood change. He is co-author of Los Angeles: The Metropolitan Experience (1976) and co-editor of Population Mobility and Residential Change (1978), and Residential Mobility and Public Policy (1980).

FRANCIS CRONIN was formerly a Senior Research Associate at the Urban Institute, Washington, D.C., and is now a member of the research staff of the Battelle Research Institute, also of Washington, D.C. While at the Urban Institute, he was involved with the analysis of the HADE (the Housing Allowance Demand Experiment) data and a number of his research papers were published by the Urban Institute.

ROBIN FLOWERDEW is Lecturer in Geography at the University of Lancaster in England. He has long been interested in issues of search in the housing market and has published several papers on housing search models.

JOHN L. GOODMAN, JR. is with the Urban Institute of Washington, D.C. and for the past five years he has been concerned with issues of mobility, housing search, and residential adjustment. He is the author of Urban Residential Mobility: Places, People, and Policy (1978) and several Urban Institute research papers.

JAMES O. HUFF is Associate Professor of Geography at the University of Illinois, Urbana, Illinois. His research interests include models of migration, particularly those concerned with small area population change. He is co-author of Christaller Central Place Structures: An Introductory Statement (1977).

KEVIN MCCARTHY is a senior economist at the Rand Corporation. In addition to his research on housing search and residential mobility, he has broad interests in population change and adjustment. His recent Rand papers include: The Changing Demographic and Economic Structure of Non-Metropolitan Areas in the 1970's (with Peter Morrison, 1978) and Housing Search and Mobility (1979).

DUNCAN MACLENNAN is at the Center for Urban and Regional Research at the University of Glasgow in Scotland and has been investigating the search behavior of both student tenants and home owners. He has recently completed a detailed study of search for furnished rental housing in the City of Glasgow. He is the author of Housing Economics: An Applied Approach (1981).

ERIC G. MOORE is Professor of Geography at Queen's University, Kingston, Ontario, and a senior researcher focusing on relocation impacts in the Community Development Strategies Evaluation project at the University of Pennsylvania. He is co-editor of The Manipulated City (1975), Population Mobility and Residential Change (1978), and Residential Mobility and Public Policy (1980).

JUN ONAKA is Assistant Professor in the School of Urban and Public Affairs at Carnegie-Mellon University in Pittsburgh, Pennsylvania. His recently completed Ph.D. in Urban Planning was focused on disequilibrium models of residential mobility.

RISA PALM is Associate Professor of Geography at the University of Colorado, Boulder. Her research interests are focused on population change, housing, and the impact of environmental hazards on decision-making. Her publications include: Urban-Social Geography from the Perspective of the Real Estate Salesman (1976) and Real Estate Agents and the Dissemination of Information on Natural Hazards in the Urban Area, Institute of Behavioral Science, Boulder, Colorado (1980).

TERRENCE R. SMITH is Associate Professor in the Department of Geography, University of California, Santa Barbara. His research interests center on individual decision-making and information processing.

GAVIN WOOD is a member of the Centre for Urban and Regional Research at the University of Glasgow in Scotland. His research interests are focused on models of housing search and information acquisition. He has published several papers in collaboration with Duncan Maclennan on information networks in local housing markets.

# INDEX